TopSolid EXERCISES

200 3D PRACTICE DRAWINGS

SACHIDANAND JHA

Dear Reader,

Thank you for choosing **TopSolid EXERCISES** book. This book is part of a family of premium-quality CADIN360 books, all of which are written by Outstanding author who combine practical experience with a gift for teaching.

CADIN360 was founded in 2016. More than 3 years later, we're still committed to producing consistently exceptional books. With each of our titles, we're working hard to set a new standard for the industry. From the paper we print on, to the authors we work with, our goal is to bring you the best books available.

I hope you see all that reflected in these pages. I'd be very interested to hear your comments and get your feedback on how we're doing. Feel free to let me know what you think about this or any other CADIN360 book by sending me an email at contactus@cadin360.com.

If you think you've found a technical error in this book, please visit https://cadin360.com/contact-us/.
Customer feedback is critical to our efforts at CADIN360.

Best regards,

Sachidanand Jha
Founder & CEO, CADIN360

TopSolid EXERCISES

Published by
CADIN360
cadin360.com

Limit of Liability/Disclaimer of Warranty:

Examination Copies

Electronic Files

Disclaimer:

Preface

TopSolid EXERCISES

- ❖ This book contain 200 CAD practice exercises and drawings.

- ❖ This book does not provide step by step tutorial to design 3D models.

- ❖ S.I Unit is used.

- ❖ Predominantly used Third Angle Projection.

- ❖ This book is for **TopSolid** and Other Feature-Based Modeling Software such as Inventor, SolidWorks, NX, Solid Edge, AutoCAD, PTC Creo etc.

- ❖ It is intended to provide Drafters, Designers and Engineers with enough 3D CAD exercises for practice on **TopSolid**.

- ❖ It includes almost all types of exercises that are necessary to provide, clear, concise and systematic information required on industrial machine part drawings.

- ❖ Third Angle Projection is intentionally used to familiarize Drafters, Designers and Engineers in Third Angle Projection to meet the expectation of world wide Engineering drawing print.

- ❖ Clear and well drafted drawing help easy understanding of the design.

- ❖ This book is for Beginner, Intermediate and Advance CAD users.

- ❖ These exercises are from Basics to Advance level.

- ❖ Each exercises can be assigned and designed separately.

- ❖ No Exercise is a prerequisite for another. All dimensions are in mm.

- ❖ Note: Assume any missing dimensions.

EX-01
Ø80
3 HOLES Ø10
DRILLED THROUGH
28
A
28 A
28
10
5
SECTION A-A
EX-02
100
20
2X R30
2X R15
120
60
30
70
20
50
20
20
100
20
80
2X R20
2X R6
30
30
20
120
P-01

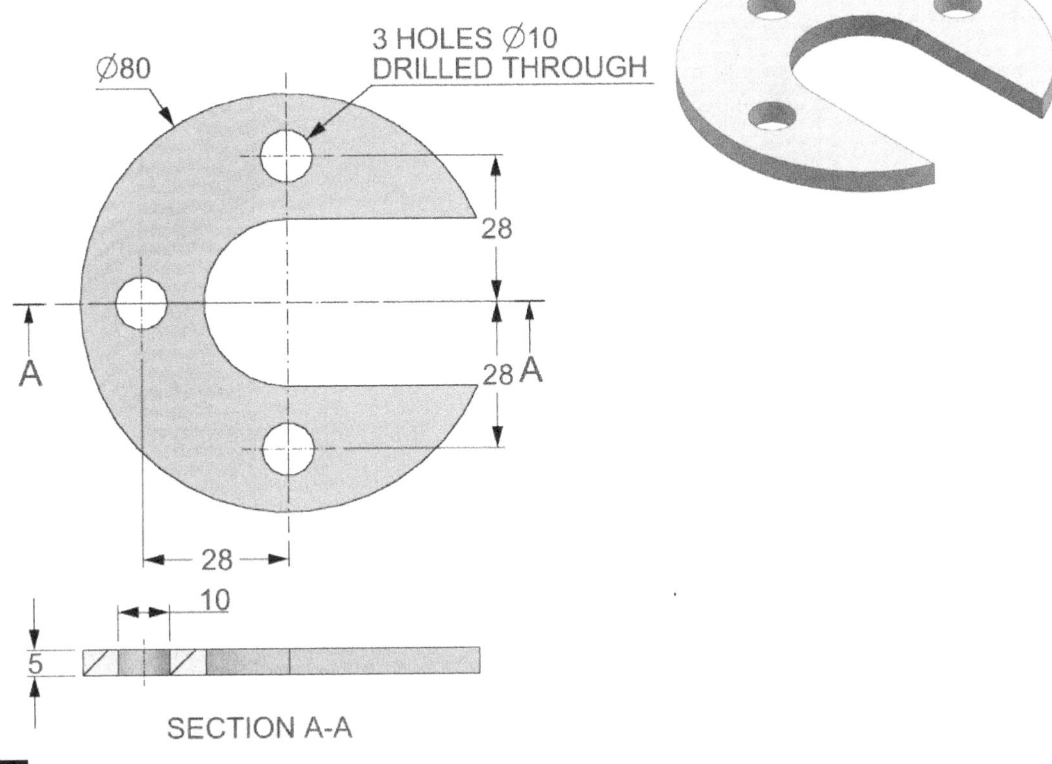

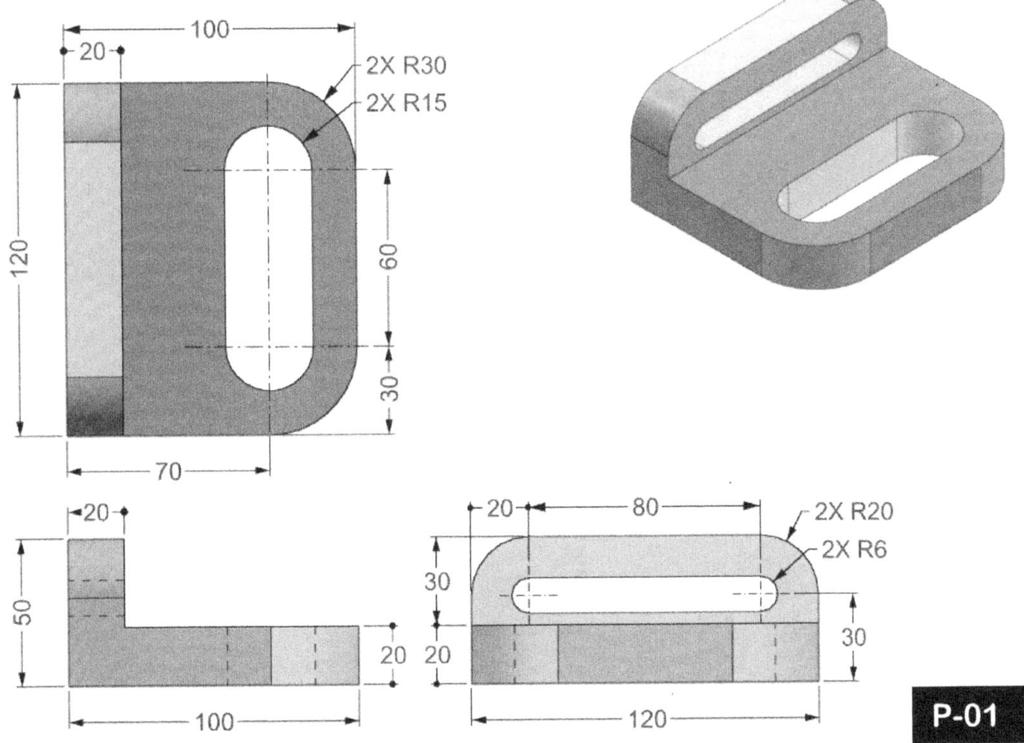

EX-03

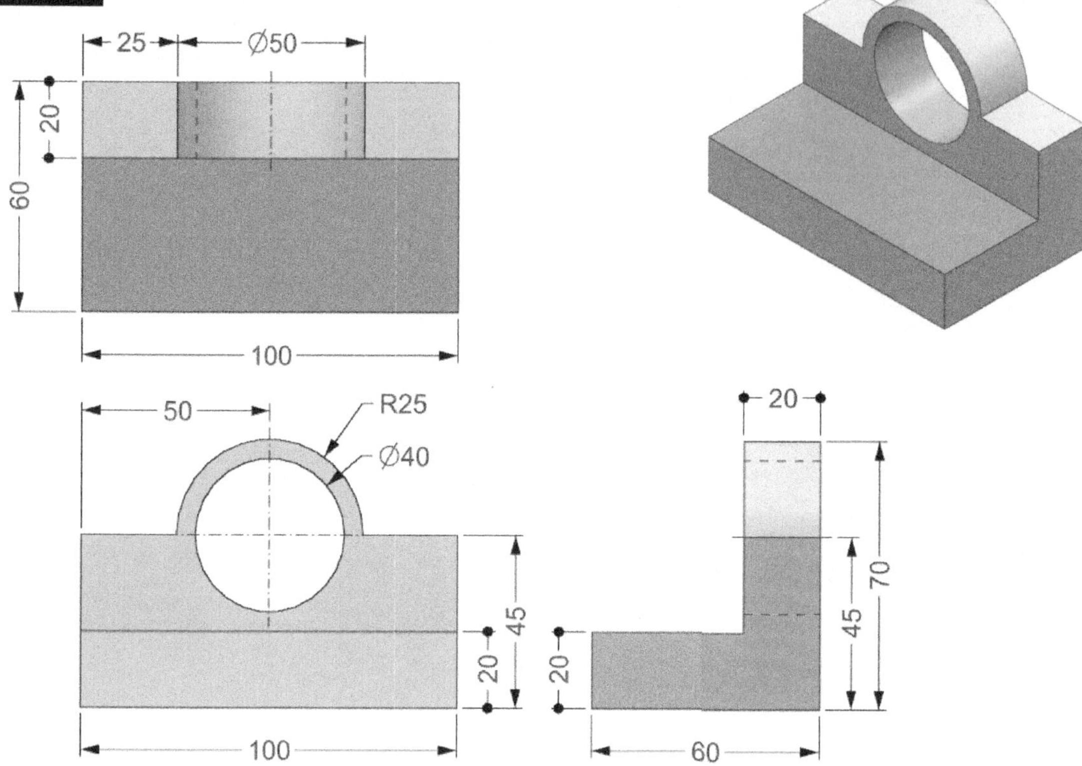
25
Ø50
20
60
100
50
R25
Ø40
45
20
100
20
70
45
20
60

EX-04

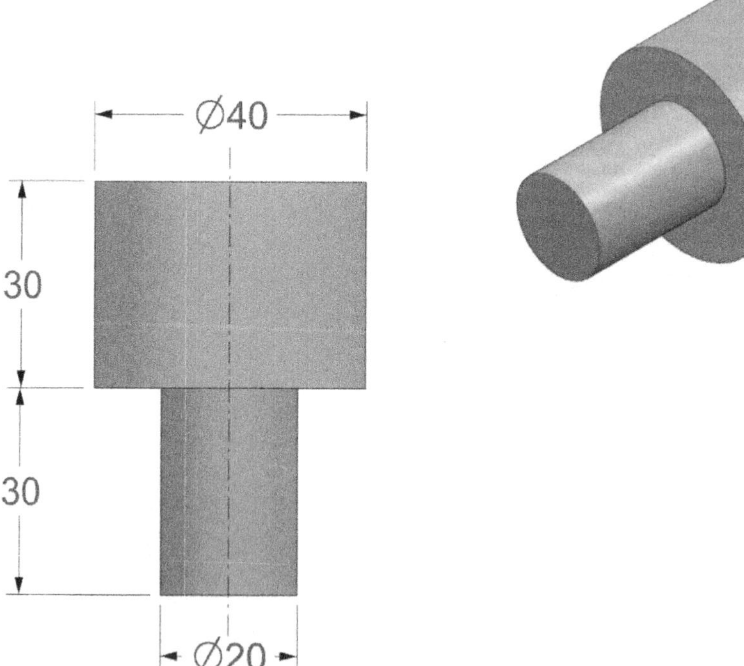
Ø40
30
30
Ø20

P-02

EX-05

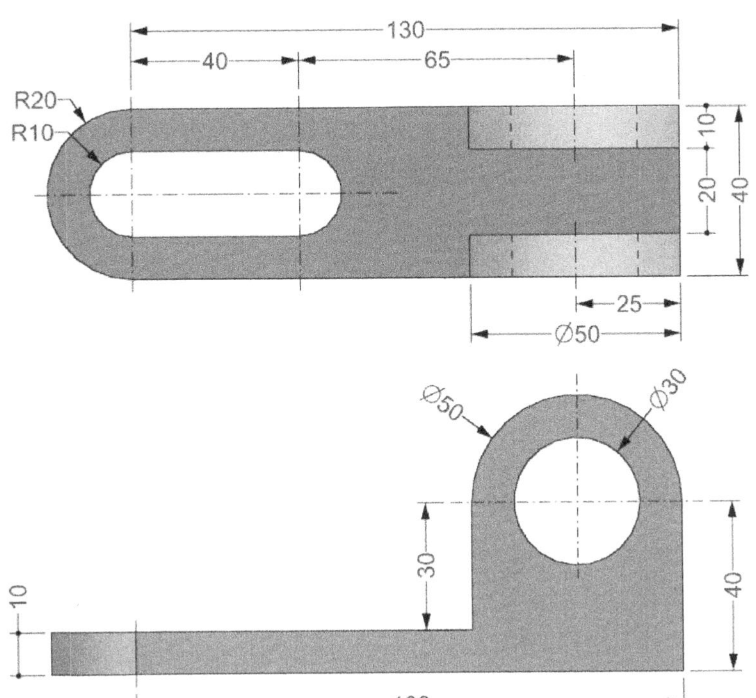
130
40
65
R20
R10
10
20
40
25
Ø50

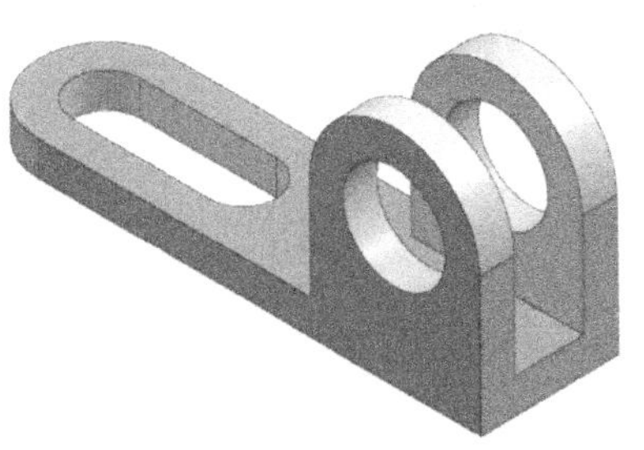

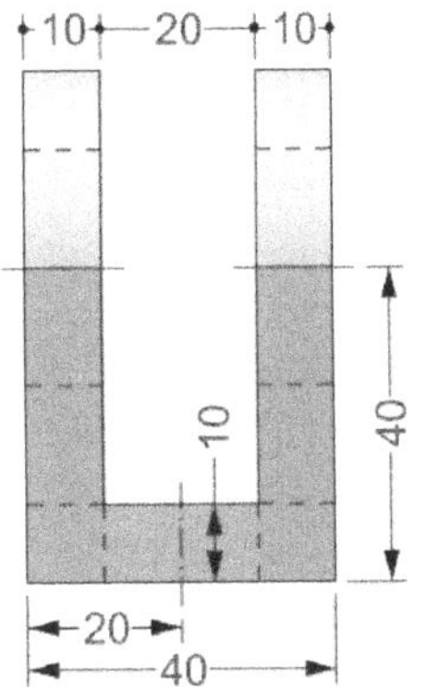
Ø50
Ø30
30
40
10
130
10
20
10
10
40
20
40

EX-06

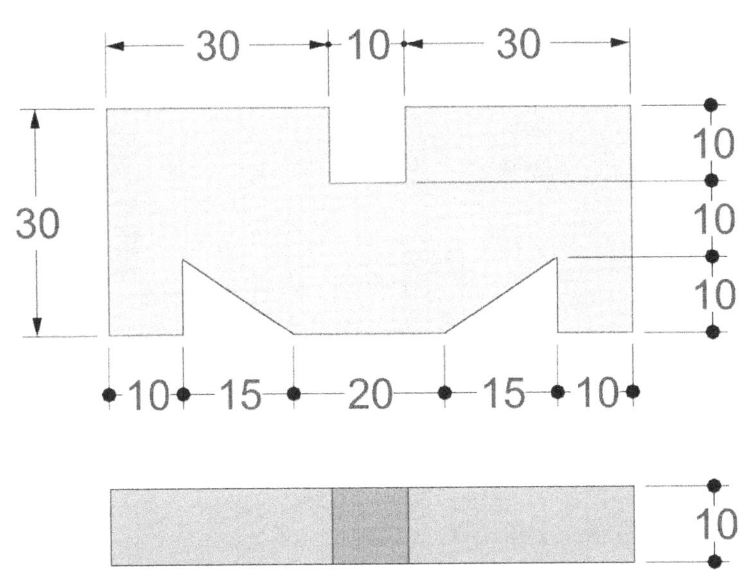
30
10
30
10
10
10
10
30
10
15
20
15
10
30
10
30
10

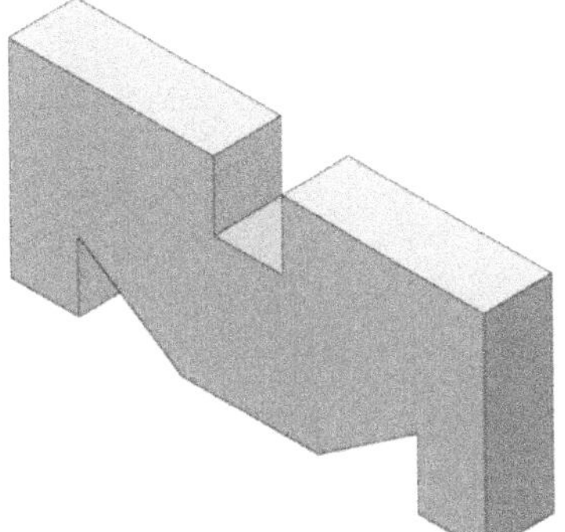

P-03

EX-07 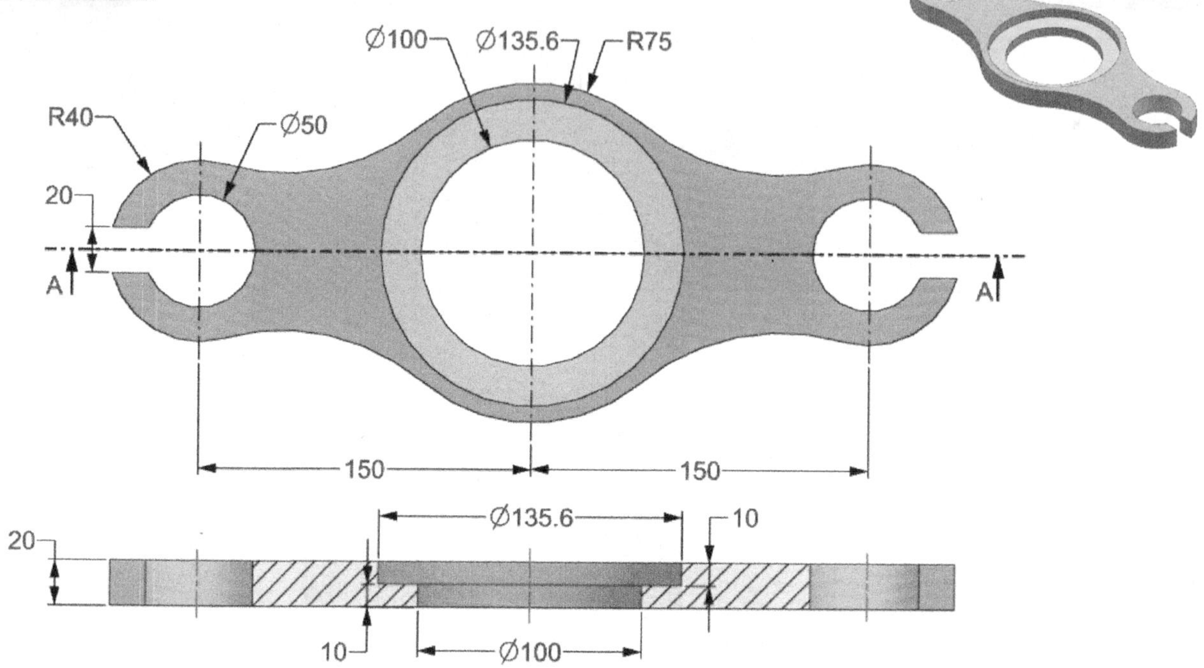
Ø100
Ø135.6
R75
R40
Ø50
20
A
A
150
150
Ø135.6
10
20
10
Ø100
SECTION A-A
(SCALE 1:1)
EX-08
50
R50
2x Ø35
2x Ø20
R25
87
5
20
45
10
45
35
20
5
10
20
P-04

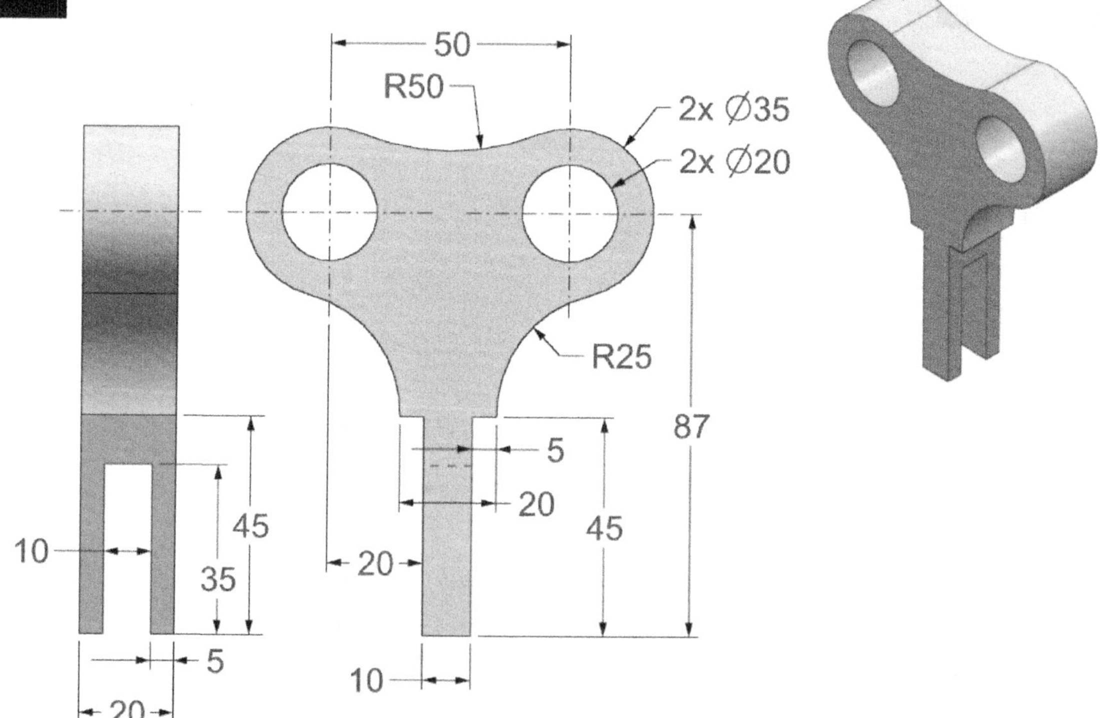

EX-09

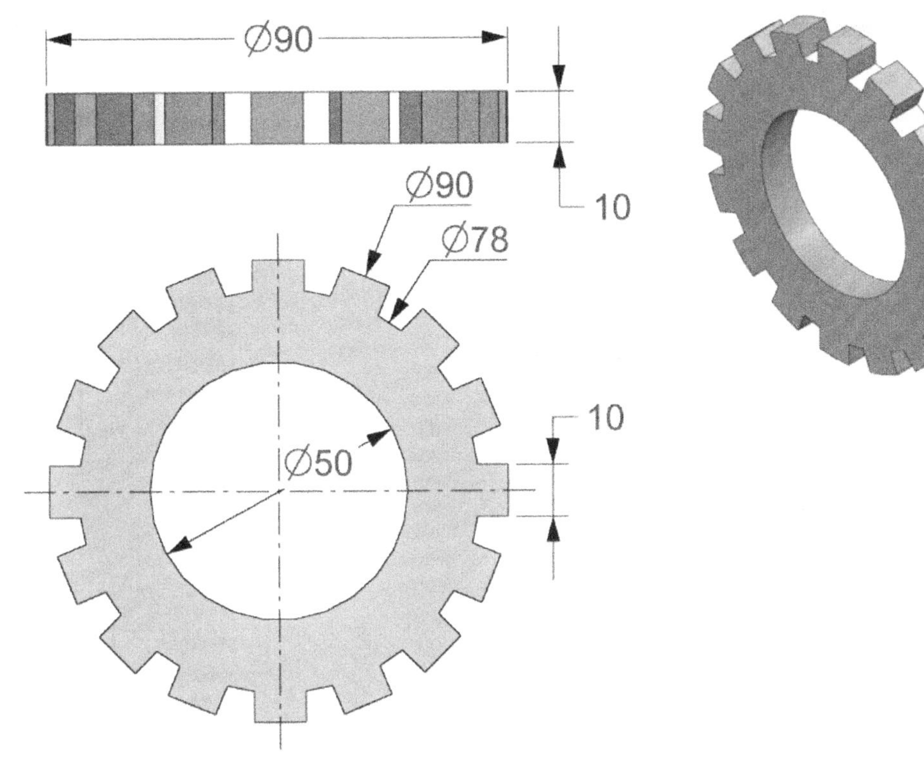
Ø90
10
Ø90
Ø78
Ø50
10

EX-10

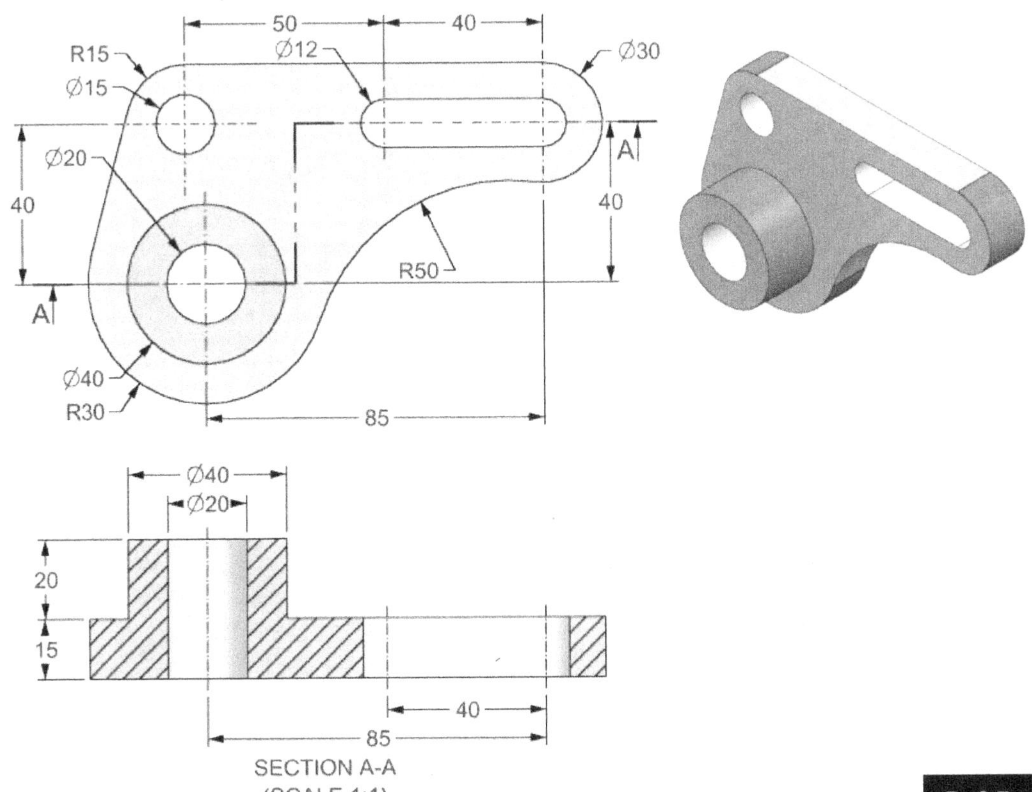
50
40
R15
Ø12
Ø30
Ø15
Ø20
A
40
40
R50
Ø40
R30
85
A
Ø40
Ø20
20
15
40
85
SECTION A-A
(SCALE 1:1)

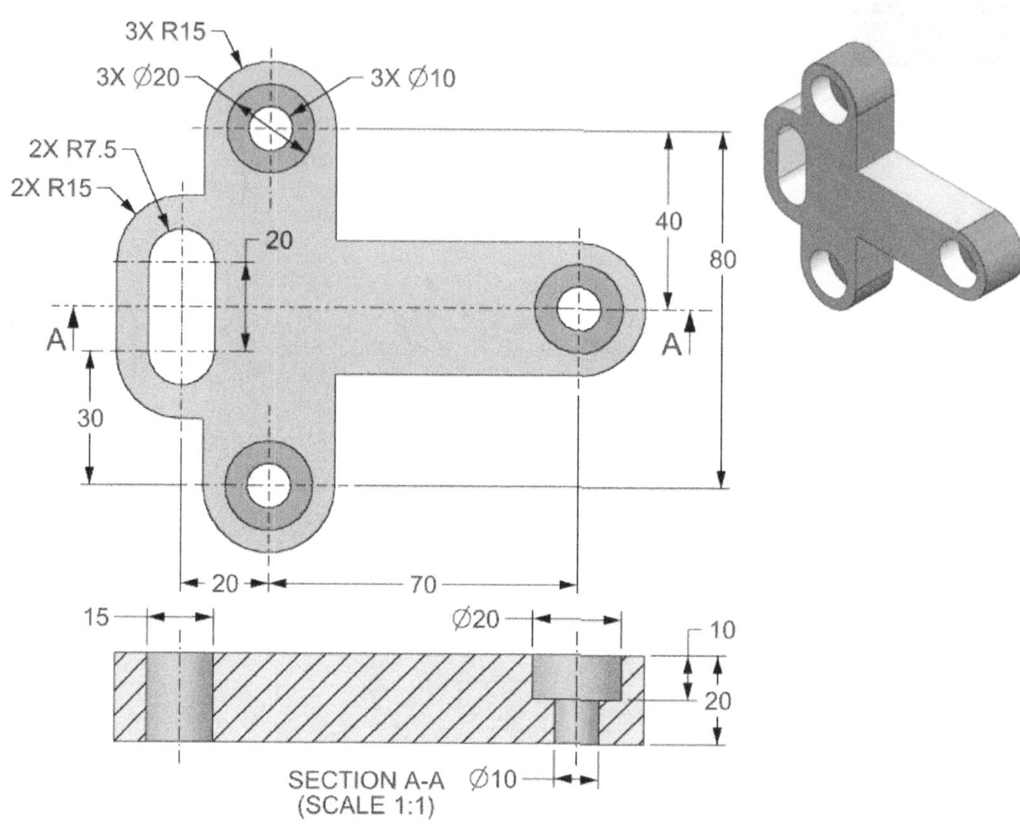

3X R15
3X Ø20
3X Ø10
2X R7.5
2X R15
20
40
80
30
20
70
15
Ø20
10
20
Ø10
SECTION A-A
(SCALE 1:1)
A
A

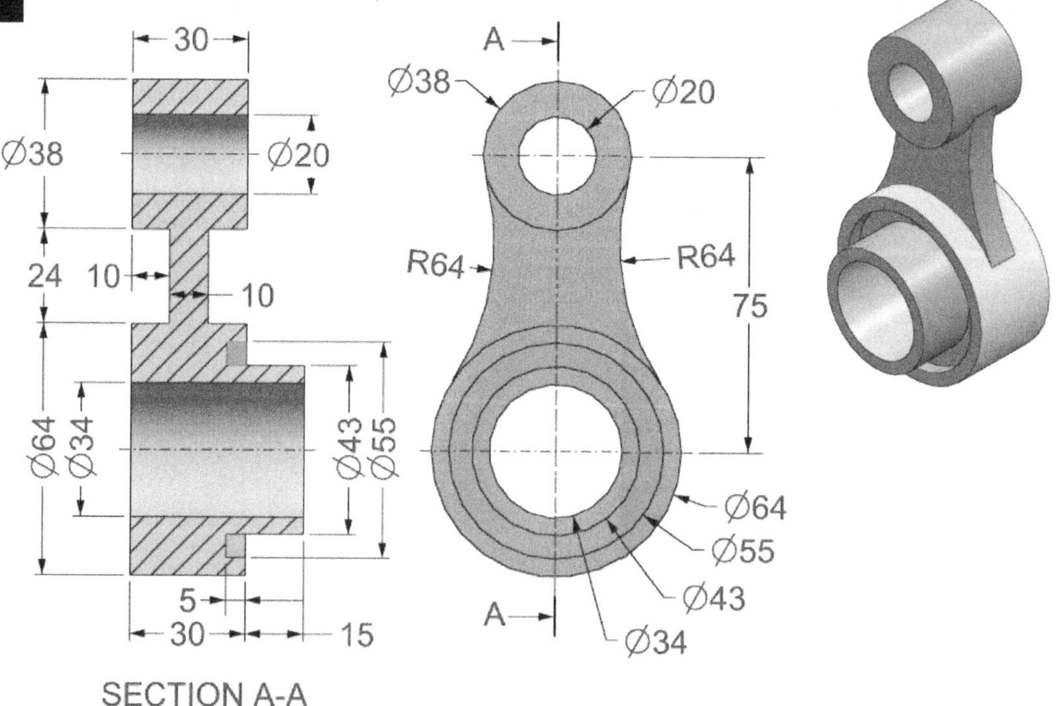

30
Ø38
Ø20
24
10
10
Ø64
Ø34
Ø43
Ø55
5
30
15
SECTION A-A
(SCALE 1:1)
A
Ø38
Ø20
R64
R64
75
Ø64
Ø55
Ø43
Ø34
A

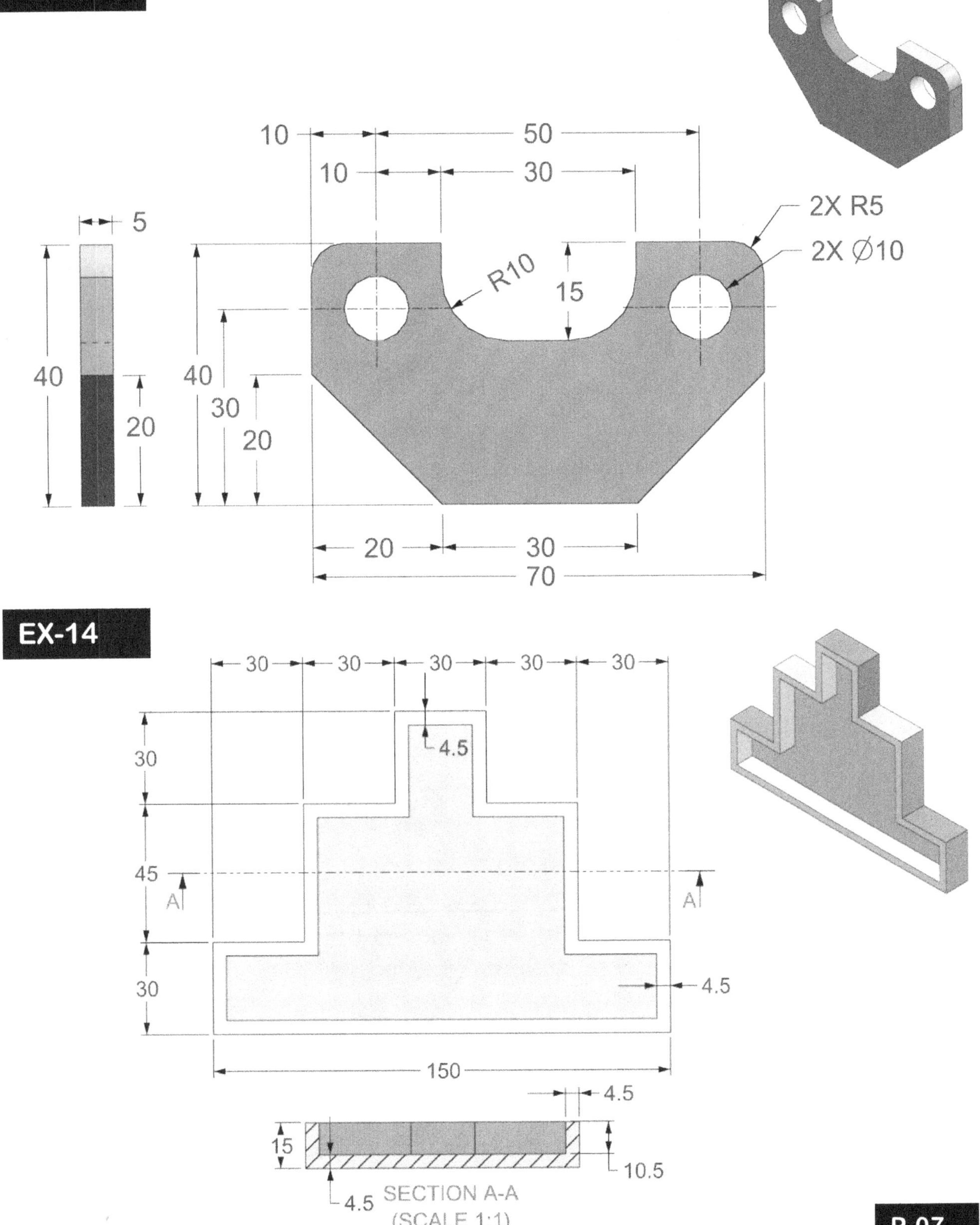

EX-13
5
40
20
10
50
10
30
2X R5
2X Ø10
R10
15
40
30
20
20
30
70
EX-14
30
30
30
30
30
30
4.5
30
45
A
A
30
4.5
150
4.5
15
10.5
4.5
SECTION A-A
(SCALE 1:1)
P-07

EX-15
4X R40
4X Ø60
4X Ø40 THRU HOLES
2X R54
10
10
200
100
100
50
50
40
2X Ø40
100
280
100
200
100
180
100
50
50
50
20
100
280
A
A
Ø60
Ø40
Ø40
5
25
25
20
280
SECTION A-A
EX-16
Ø50
Ø34
A
13.8
R32
20
40
90
26
A
164.9
10
Ø34
32
(SCALE 1:1) SECTION A-A
P-08

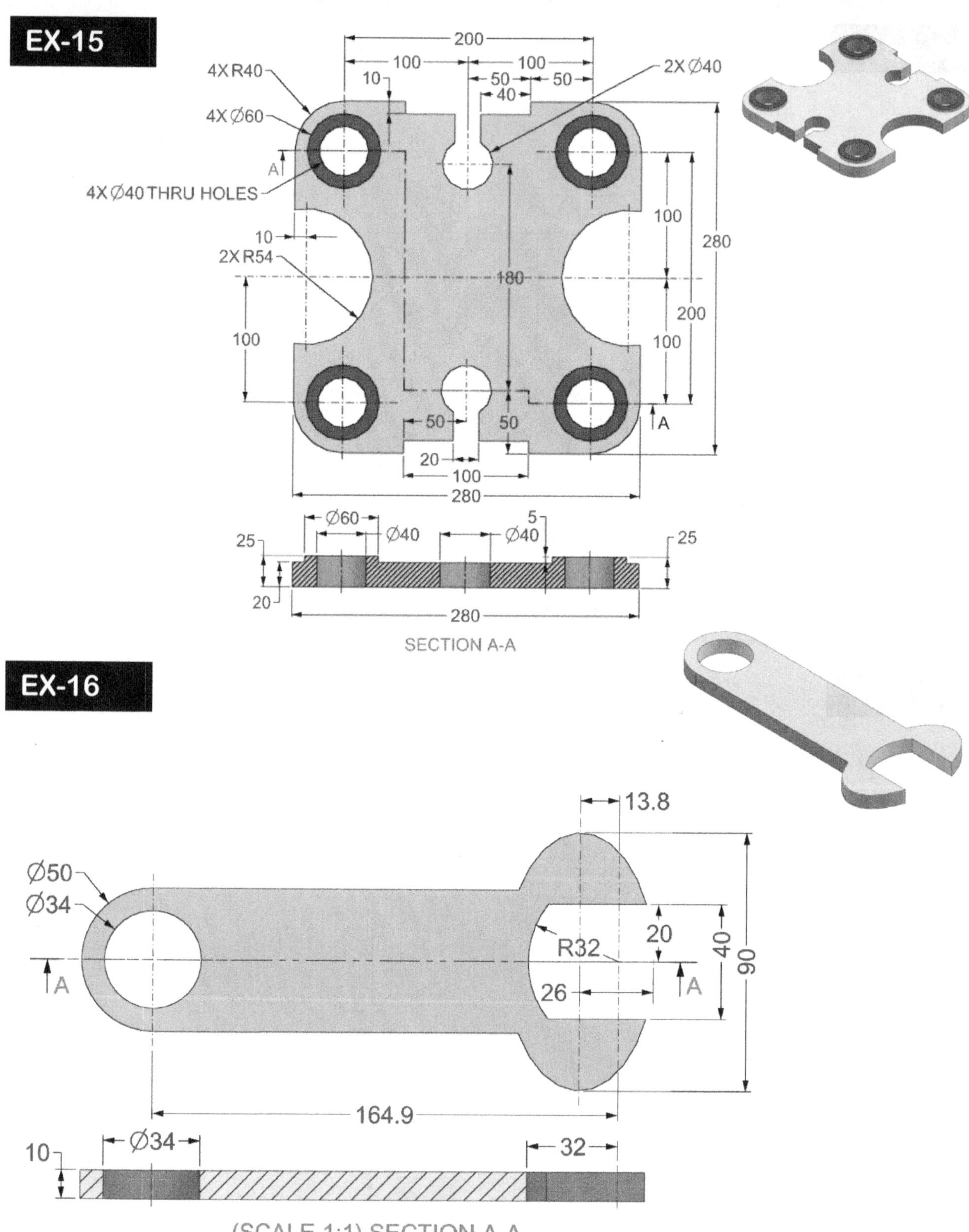

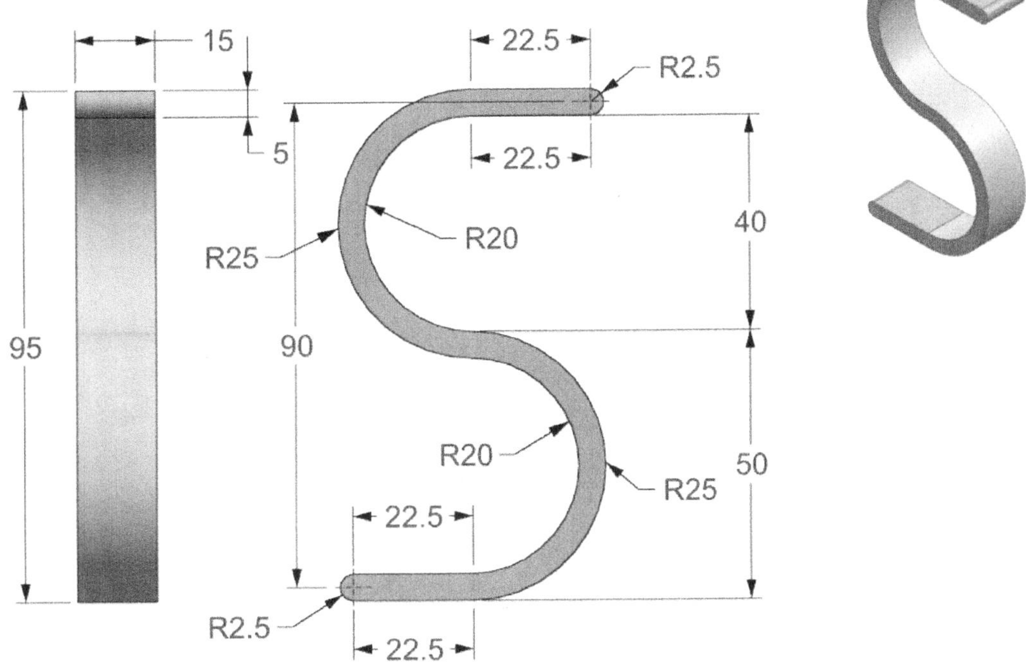

15
22.5
R2.5
5
22.5
R25
R20
40
95
90
R20
50
R25
22.5
R2.5
22.5

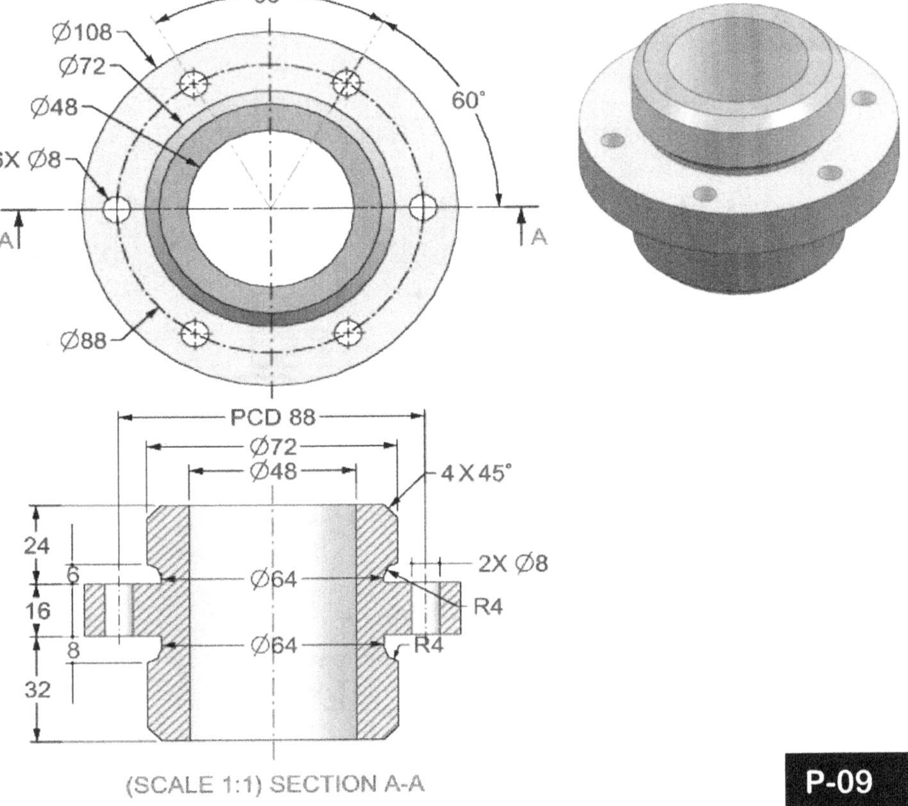

60°
Ø108
Ø72
Ø48
60°
6X Ø8
A
A
Ø88
PCD 88
Ø72
Ø48
4 X 45°
24
6
2X Ø8
Ø64
16
R4
8
Ø64
R4
32
(SCALE 1:1) SECTION A-A

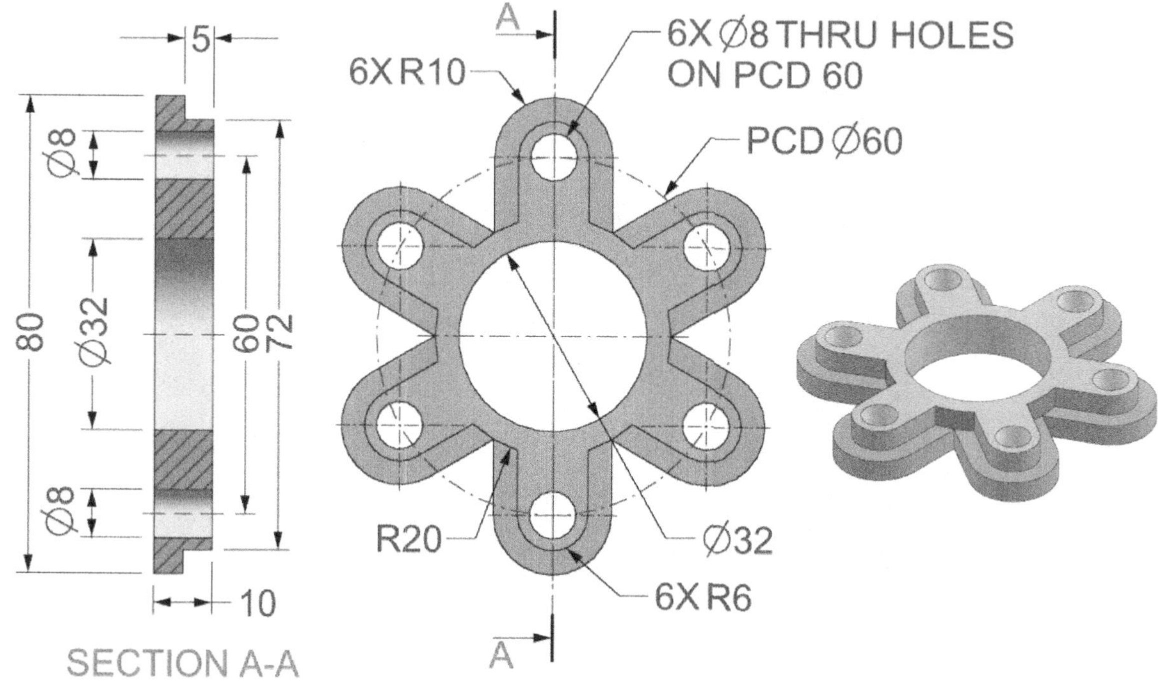

EX-19

5
Ø8
Ø32
80
60
72
Ø8
10
SECTION A-A
A
6X R10
6X Ø8 THRU HOLES
ON PCD 60
PCD Ø60
R20
Ø32
6X R6
A

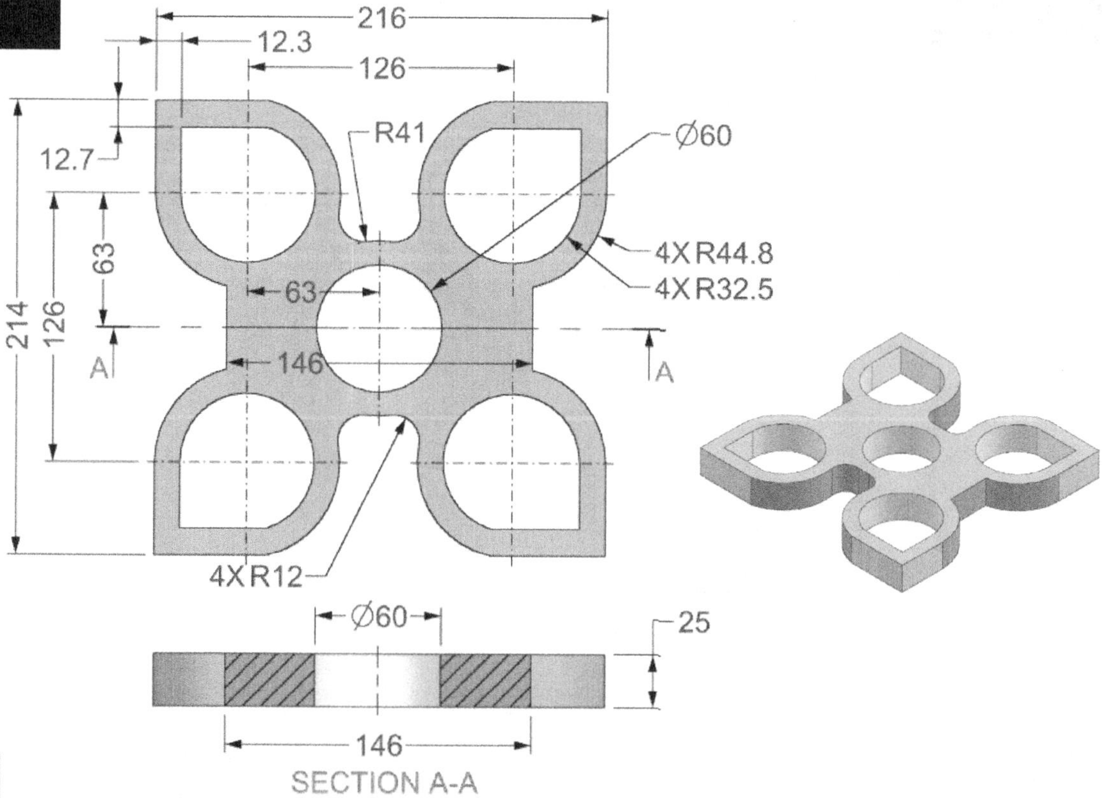

EX-20

216
12.3
126
12.7
R41
Ø60
63
63
214
126
4X R44.8
4X R32.5
A
146
A
4X R12
Ø60
25
146
SECTION A-A

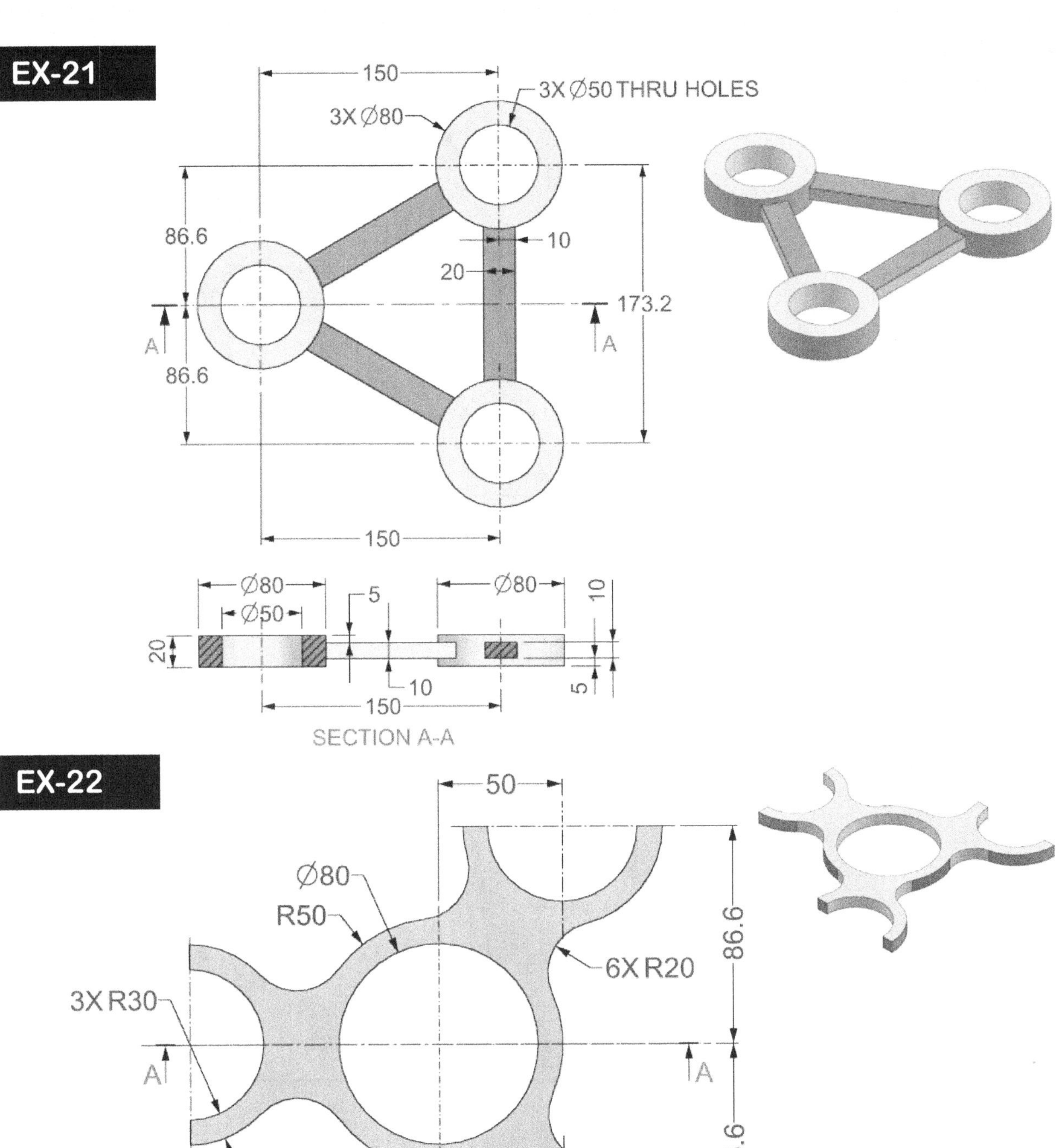

EX-21
150
3X Ø80
3X Ø50 THRU HOLES
86.6
10
20
173.2
A
A
86.6
150
Ø80
Ø50
5
Ø80
10
20
10
5
150
SECTION A-A
EX-22
50
Ø80
R50
86.6
6X R20
3X R30
A
A
3X R40
86.6
100
50
10
Ø80
SECTION A-A
P-11

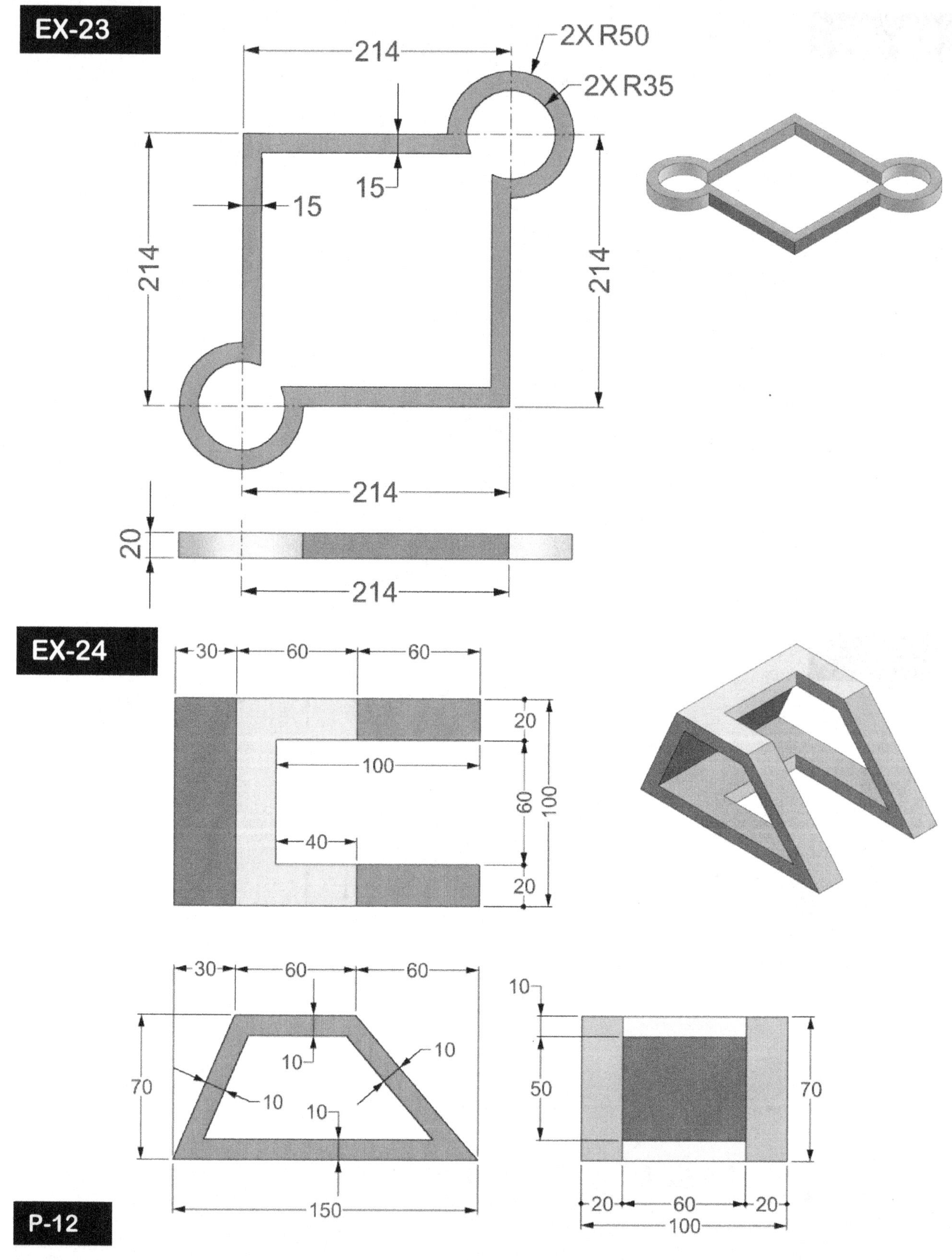

EX-23
214
2X R50
2X R35
15
15
15
214
214
214
214
20
214
EX-24
30
60
60
20
100
100
60
40
20
30
60
60
10
10
70
10
10
10
150
10
50
70
20
60
20
100
P-12

EX-25 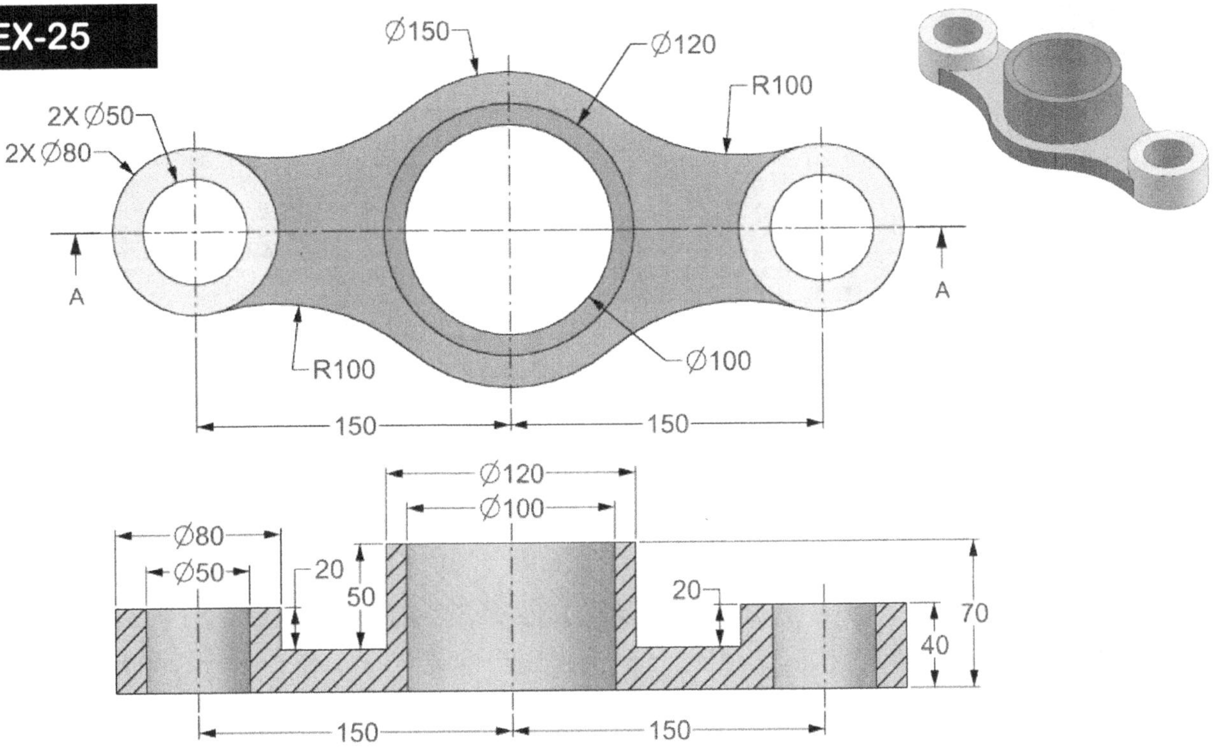
Ø150
Ø120
R100
2X Ø50
2X Ø80
R100
Ø100
150
150
Ø120
Ø100
Ø80
Ø50
20
50
20
70
40
150
150
SECTION A-A
(SCALE 1:1)
EX-26
Ø24
Ø44
Ø36
A
A
Ø44
Ø36
Ø32
Ø24
2X45°
12
4
8
36
3
12
SECTION A-A
(SCALE 1:1)
P-13

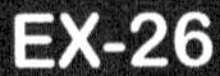
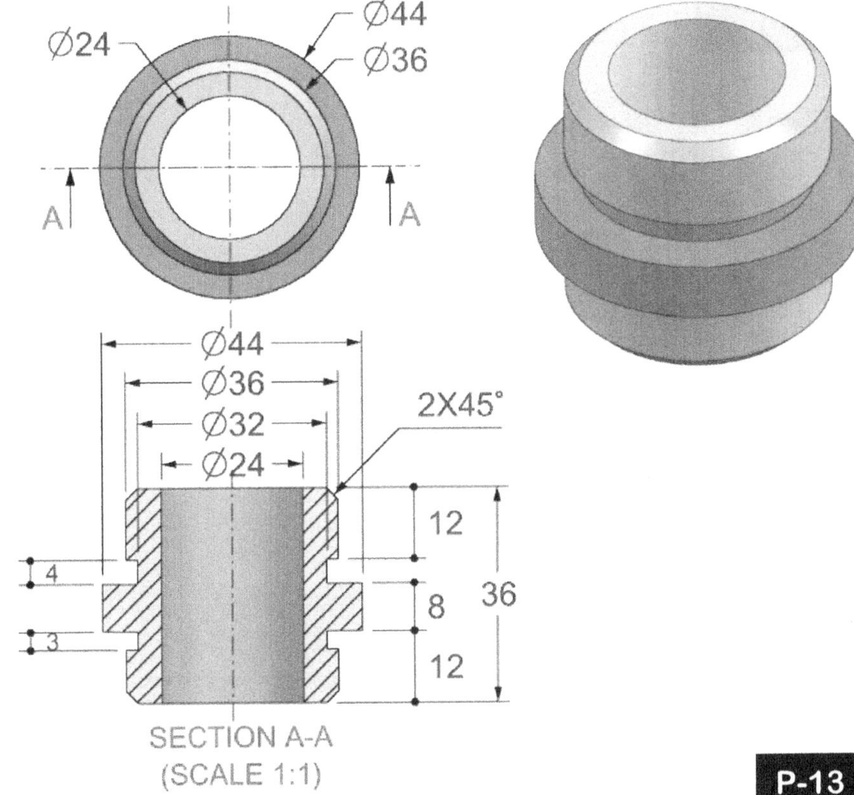

EX-27 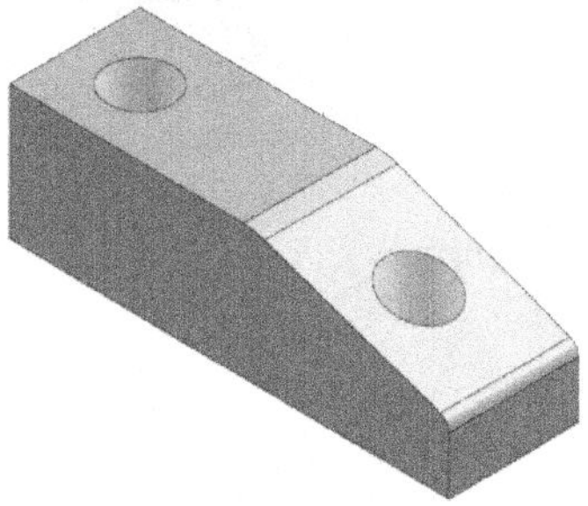 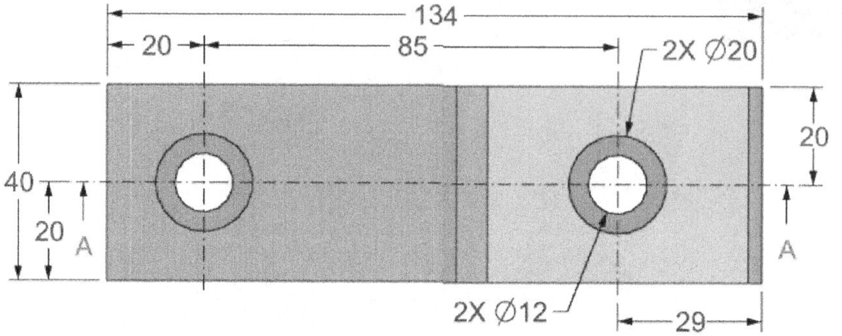 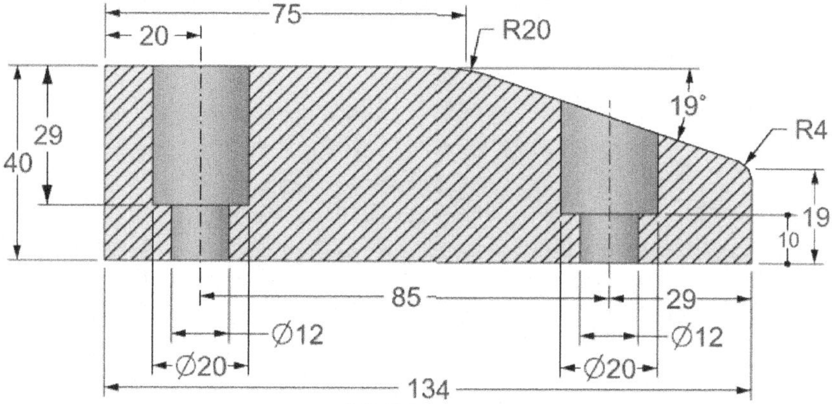
134
20
85
2X Ø20
40
20
20
A
A
2X Ø12
29
75
20
R20
19°
29
40
R4
19
10
85
29
Ø12
Ø12
Ø20
Ø20
134
SECTION A-A
(SCALE 1:1)
EX-28 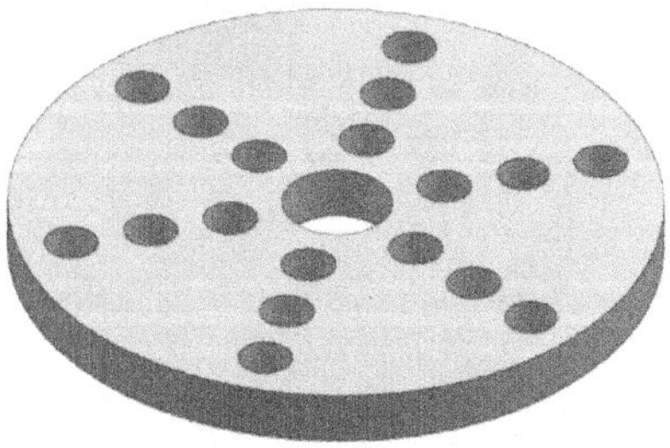 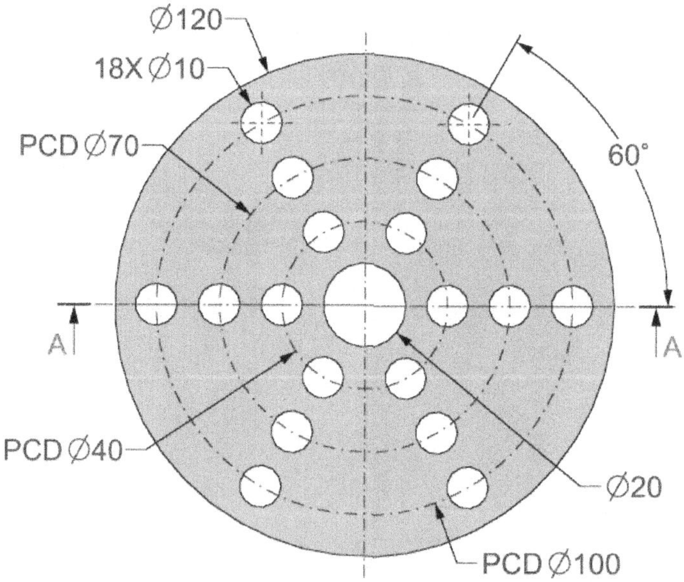 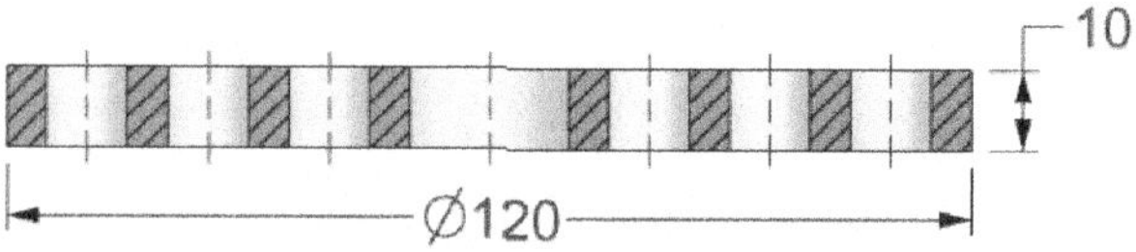
Ø120
18X Ø10
60°
PCD Ø70
A
A
PCD Ø40
Ø20
PCD Ø100
10
Ø120
SECTION A-A
P-14

EX-29

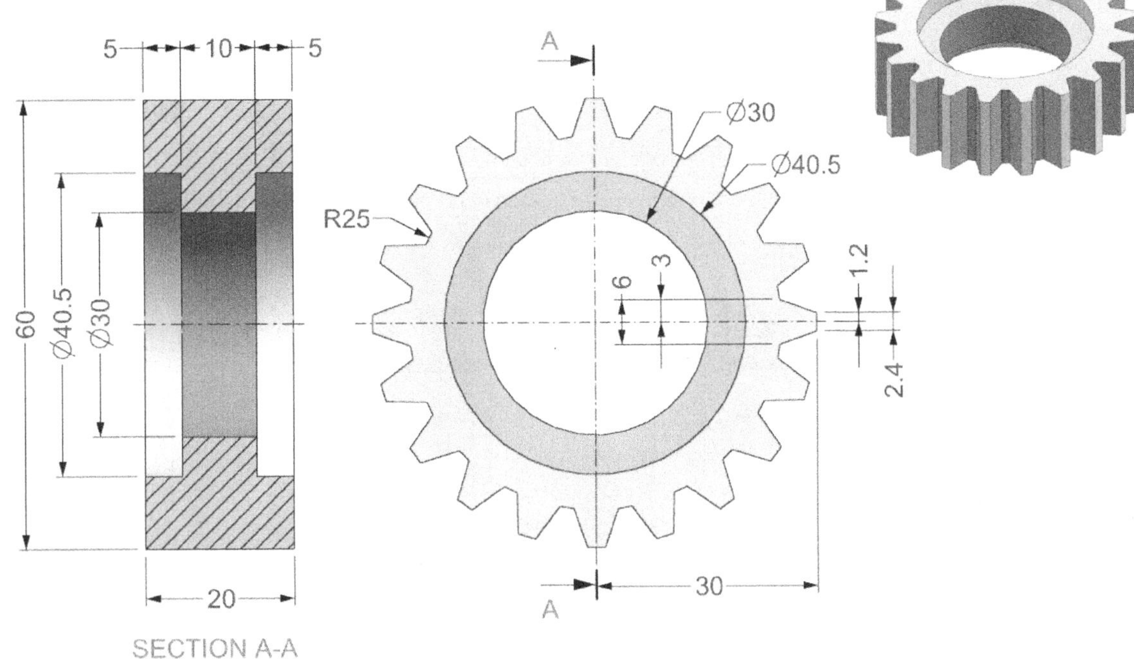

5
10
5
A
Ø30
Ø40.5
R25
6
3
1.2
2.4
Ø40.5
Ø30
60
20
30
A
SECTION A-A

EX-30 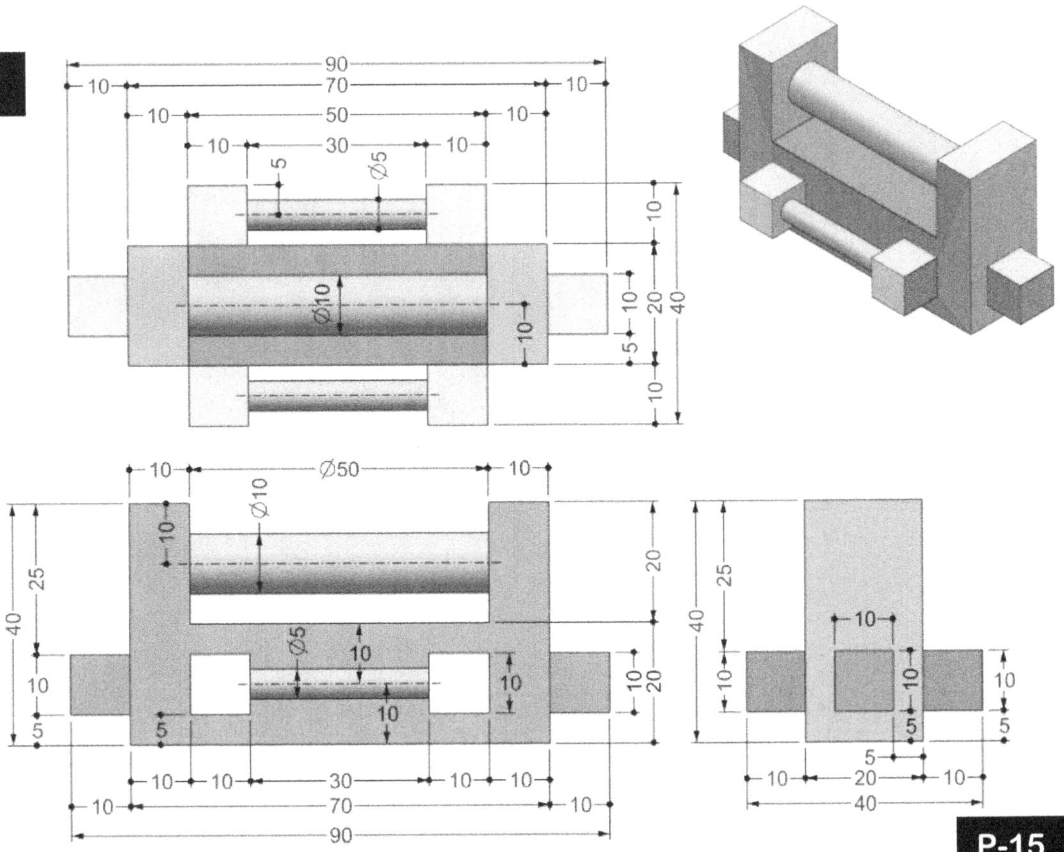
90
70
10
10
50
10
10
30
10
10
5
Ø5
Ø10
10
20
40
10
5
10
Ø50
Ø10
10
25
40
Ø5
10
10
10
20
20
10
10
10
30
10
10
5
5
10
10
70
10
10
90
25
40
10
10
10
5
5
10
20
10
10
40
P-15

EX-31
40
Ø20
Ø20
Ø16
R5
R2
R2
40
60
100
SECTION A-A
60
20
40
Ø40
Ø40
Ø20
A
A
Ø20
Ø16
10
15
60
20
Ø20
R2
R5
40
60
20
40
30
R2
40
10
15
20
10
15

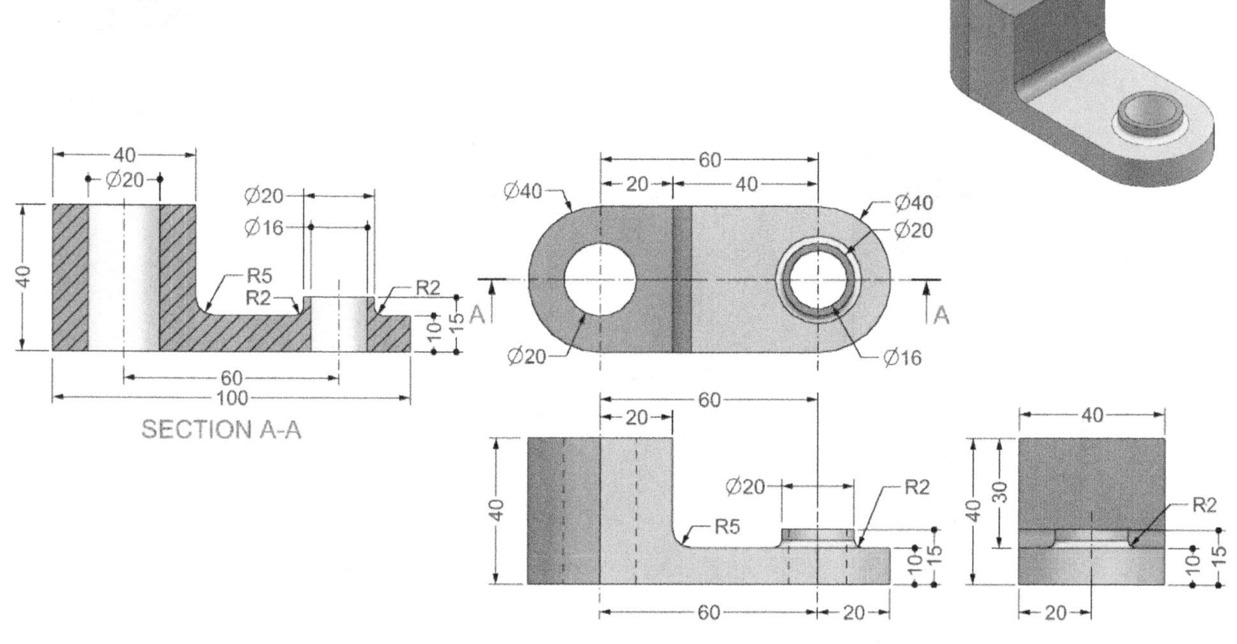

EX-32
3X Ø30 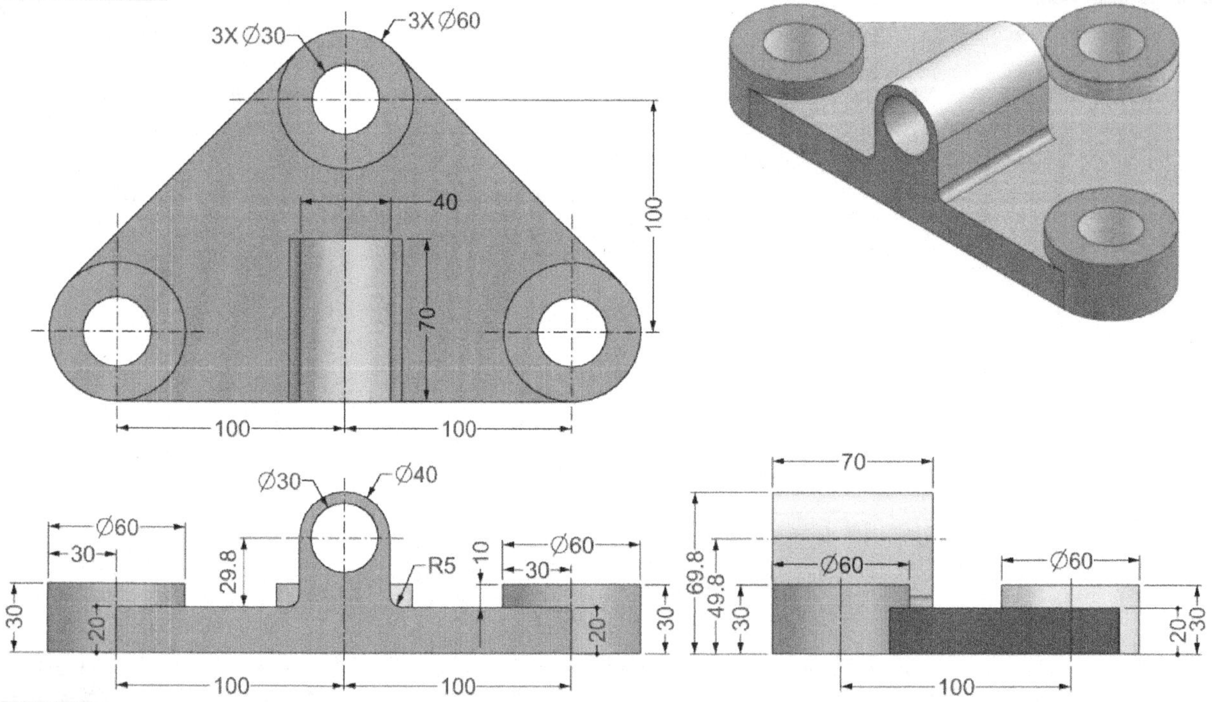
3X Ø60
40
70
100
100
100
Ø30
Ø40
Ø60
30
29.8
R5
10
30
Ø60
30
20
100
100
20
70
69.8
49.8
Ø60
30
Ø60
20
30
100
P-16

EX-33

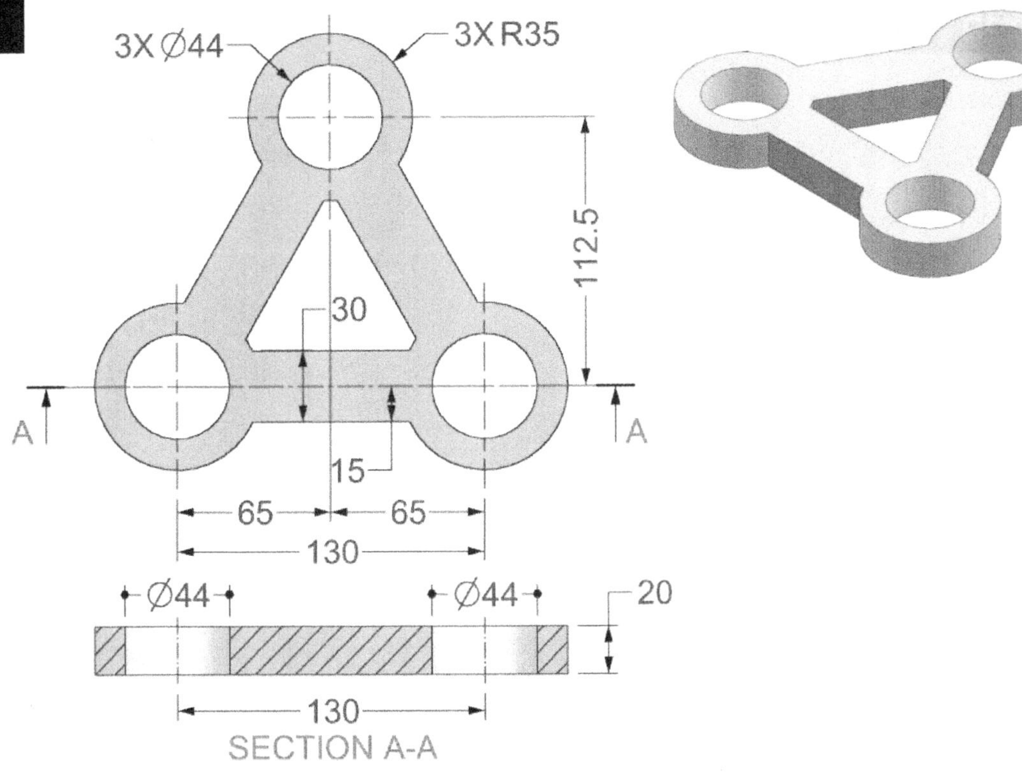
3X Ø44
3X R35
112.5
30
15
A
A
65
65
130
Ø44
Ø44
20
130
SECTION A-A

EX-34

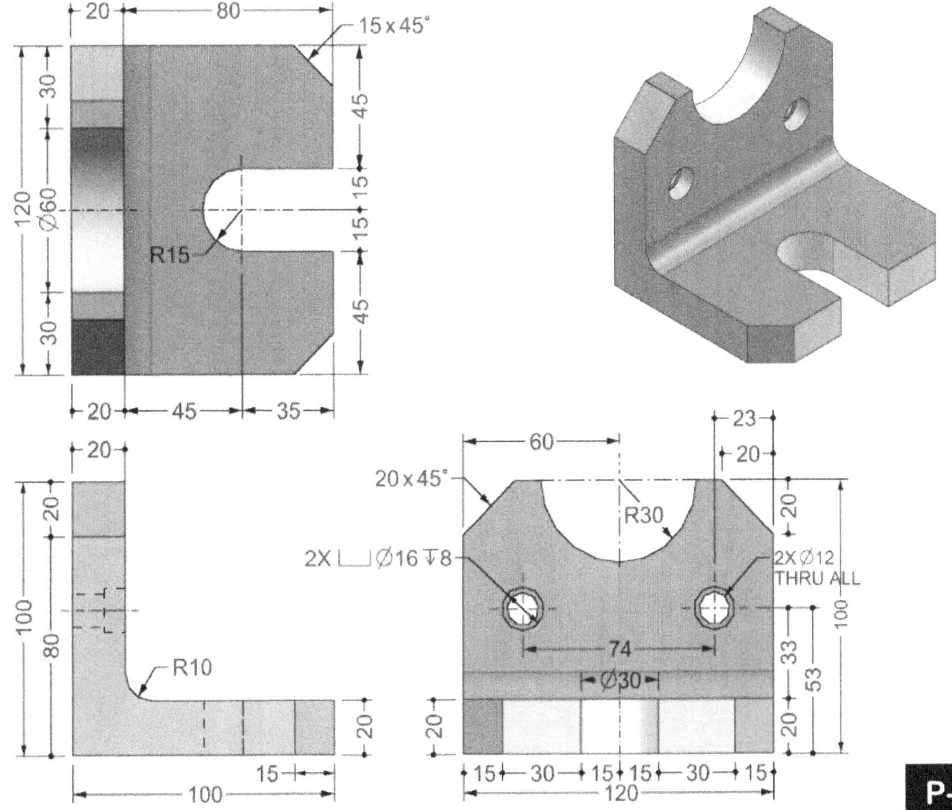
20
80
15 x 45°
30
45
120
Ø60
15 15
15
R15
45
30
20
45
35
20
20
100
80
R10
15
100
23
60
20
20 x 45°
R30
20
2X ⌴ Ø16 ↧8
2X Ø12
THRU ALL
74
33
100
Ø30
53
20
20
15 30 15 15 30 15
120

P-17

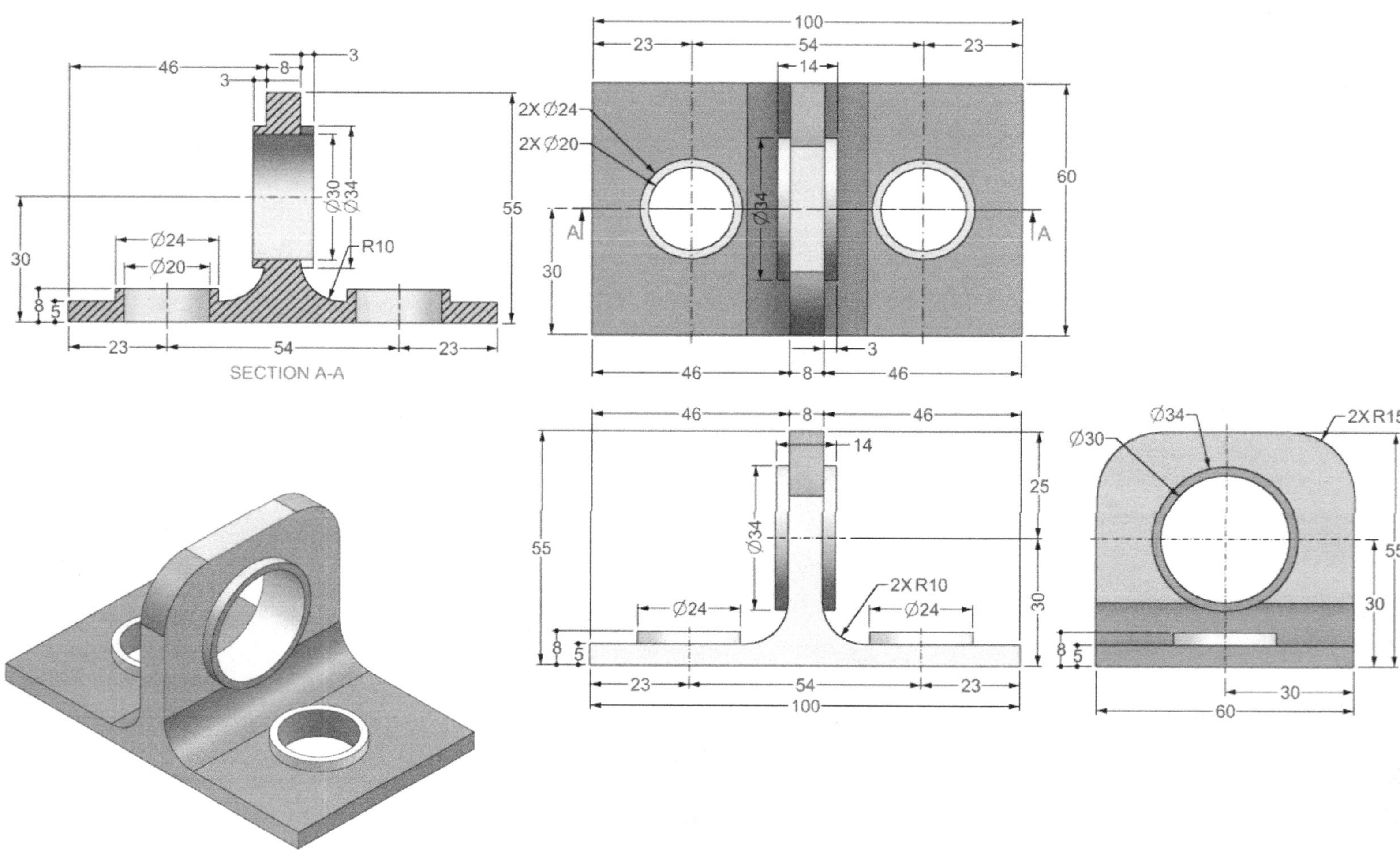

46
3
8
3
Ø30
Ø34
55
30
Ø24
Ø20
R10
8
5
23
54
23
SECTION A-A
100
23
54
23
14
2X Ø24
2X Ø20
Ø34
60
30
A
A
46
8
46
3
46
8
46
14
Ø34
25
55
30
Ø24
Ø24
2X R10
8
5
23
54
23
100
Ø30
Ø34
2X R15
55
30
8
5
30
60

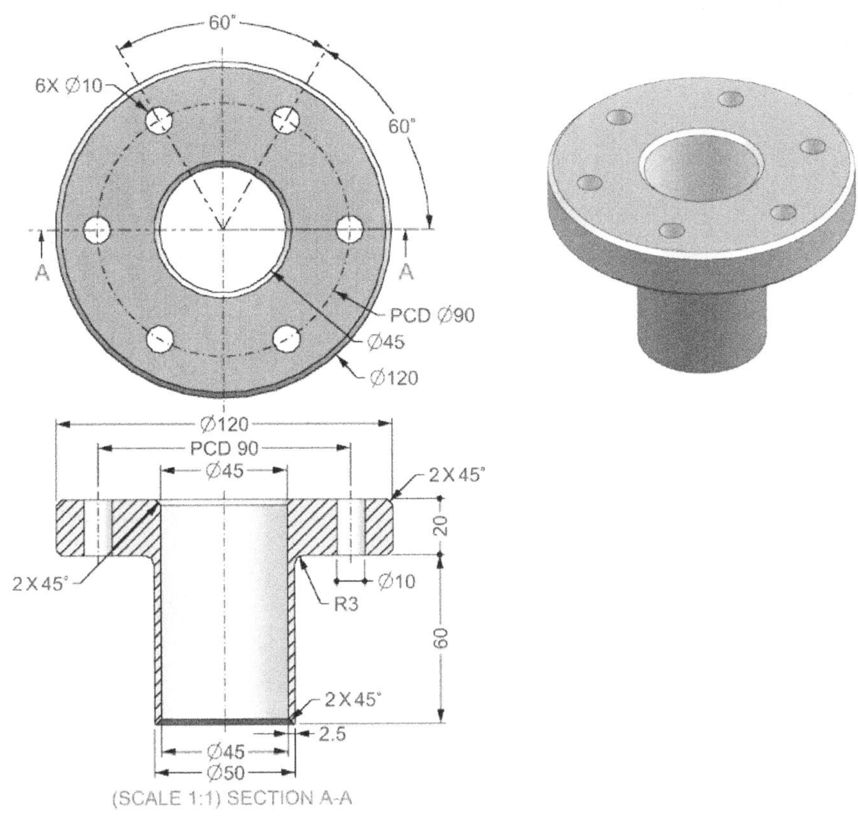

EX-36

60°
6X Ø10
60°
A
A
PCD Ø90
Ø45
Ø120
Ø120
PCD 90
Ø45
2 X 45°
20
2 X 45°
Ø10
R3
60
2 X 45°
2.5
Ø45
Ø50

(SCALE 1:1) SECTION A-A

EX-37

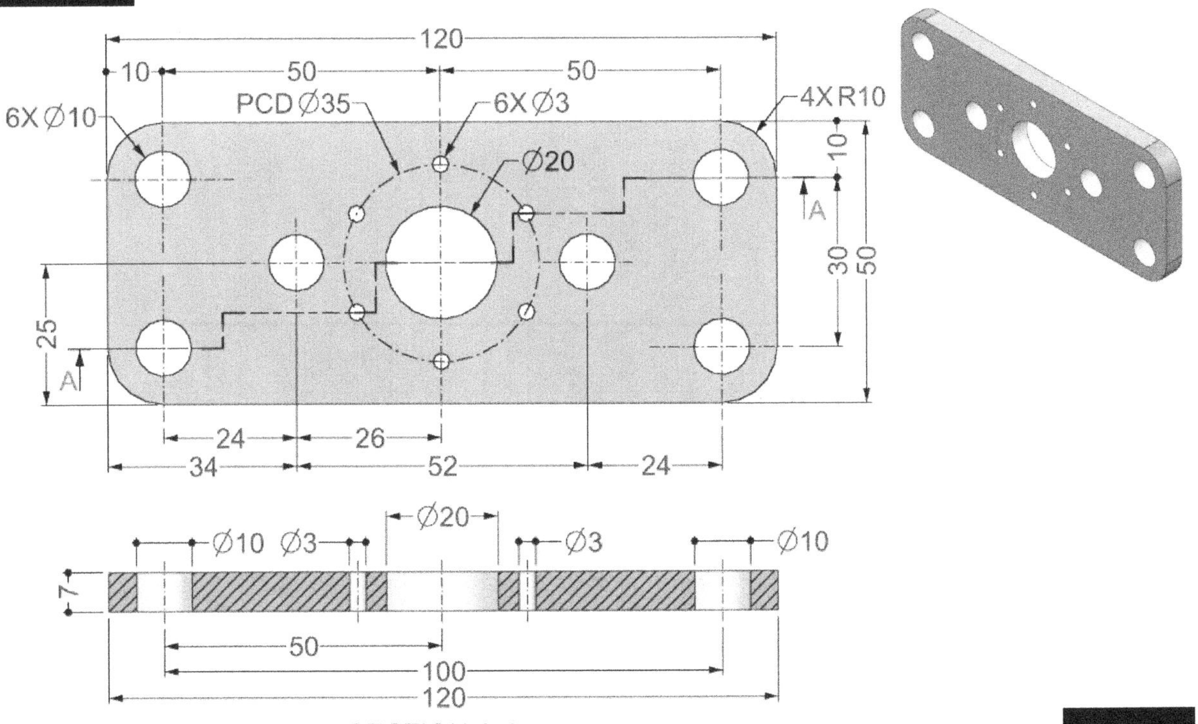

120
10
50
50
6X Ø10
PCD Ø35
6X Ø3
4X R10
10
Ø20
A
30
50
25
A
24
26
34
52
24
Ø20
Ø10
Ø3
Ø3
Ø10
7
50
100
120

SECTION A-A

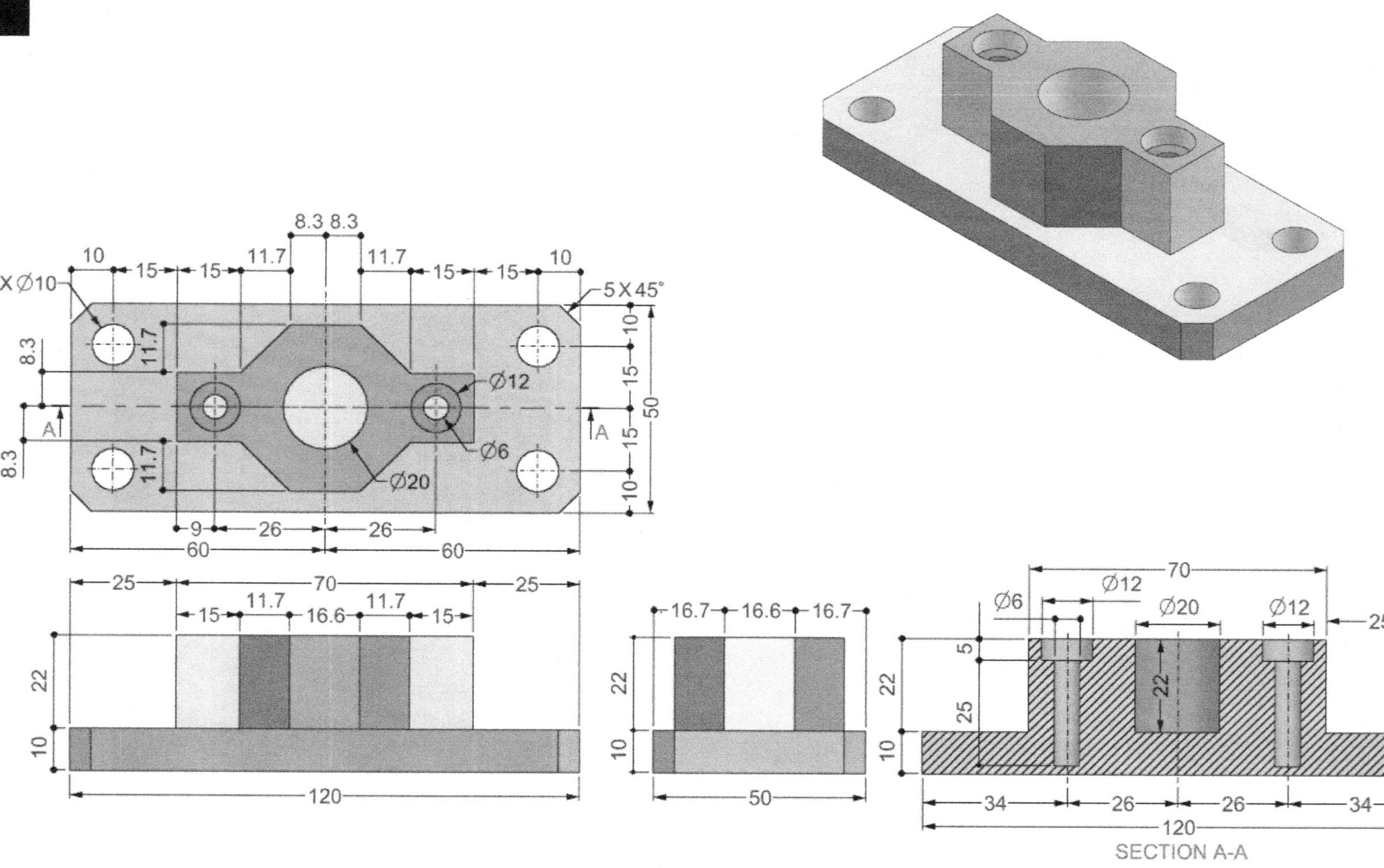
4X Ø10
5 X 45°
8.3 8.3
10
15
15
11.7
11.7
15
15
10
11.7
11.7
8.3
8.3
8.3
Ø12
Ø6
Ø20
A
A
10
15
15
10
50
9
26
26
60
60
25
15
11.7
16.6
11.7
15
25
70
22
10
120
16.7
16.6
16.7
22
10
50
Ø6
Ø12
Ø20
Ø12
70
5
25
22
22
10
34
26
26
34
120
SECTION A-A

EX-39
70 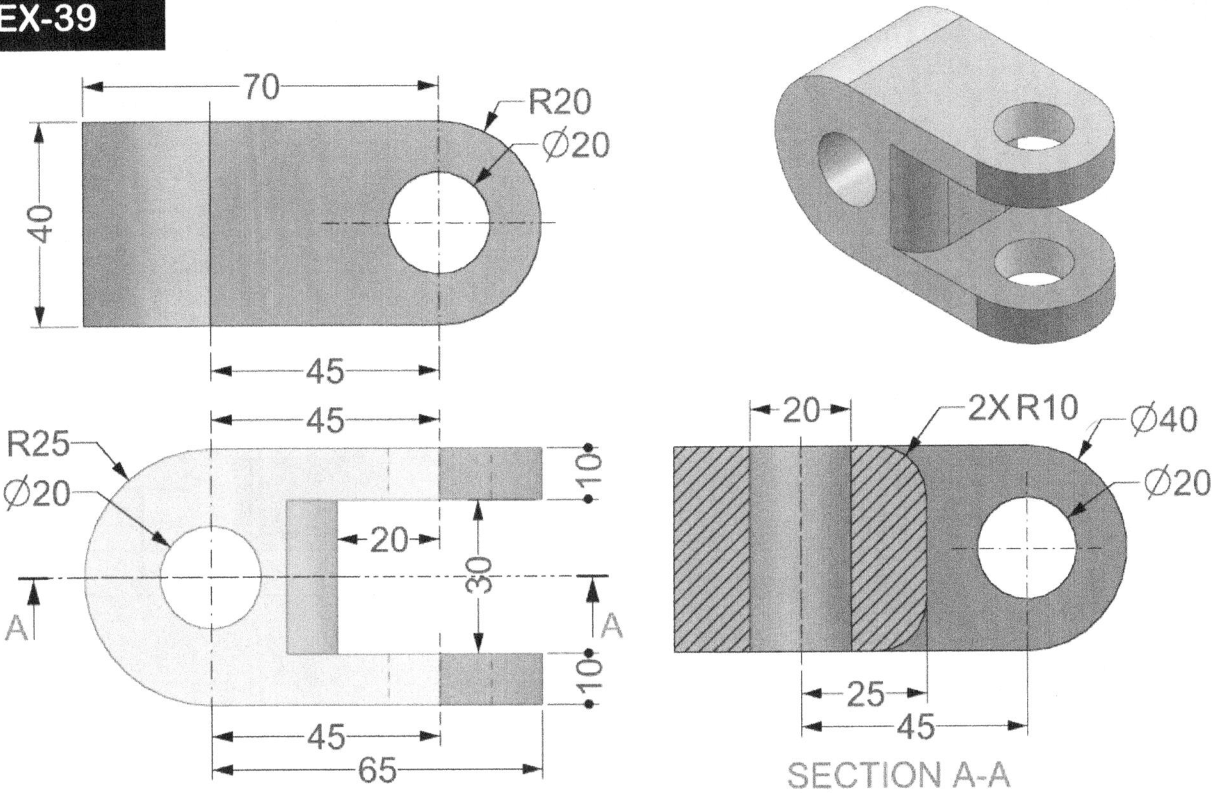
R20
Ø20
40
45
R25
Ø20
45
10
20
30
10
A
A
45
65
20
2X R10
Ø40
Ø20
25
45
SECTION A-A
EX-40
Ø60 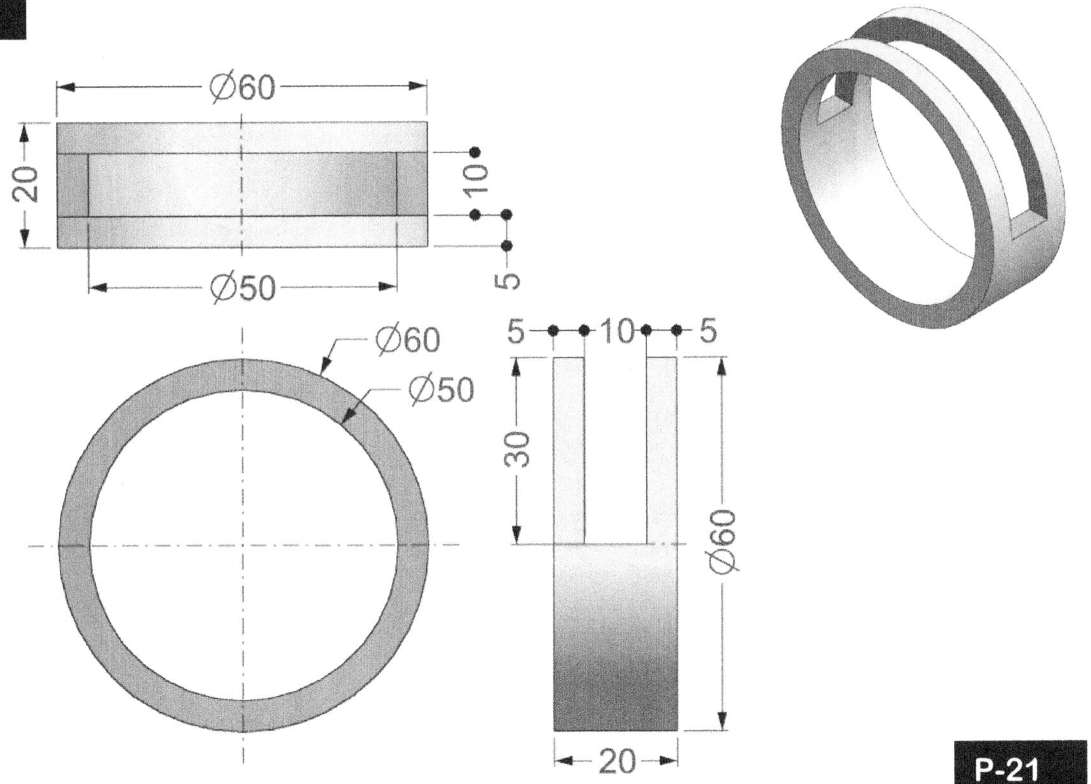
20
10
Ø50
5
Ø60
Ø50
5
10
5
30
Ø60
20
P-21

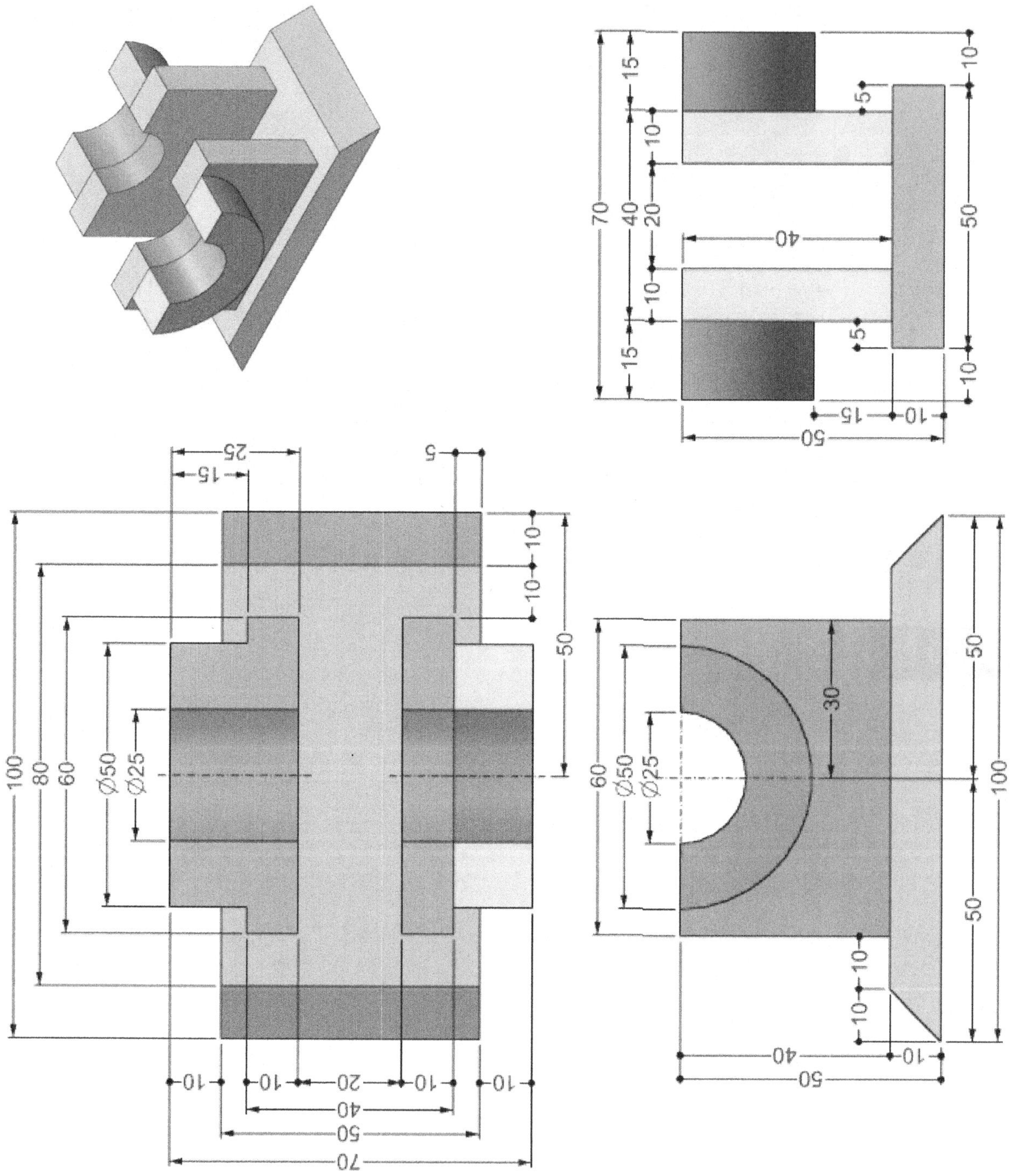

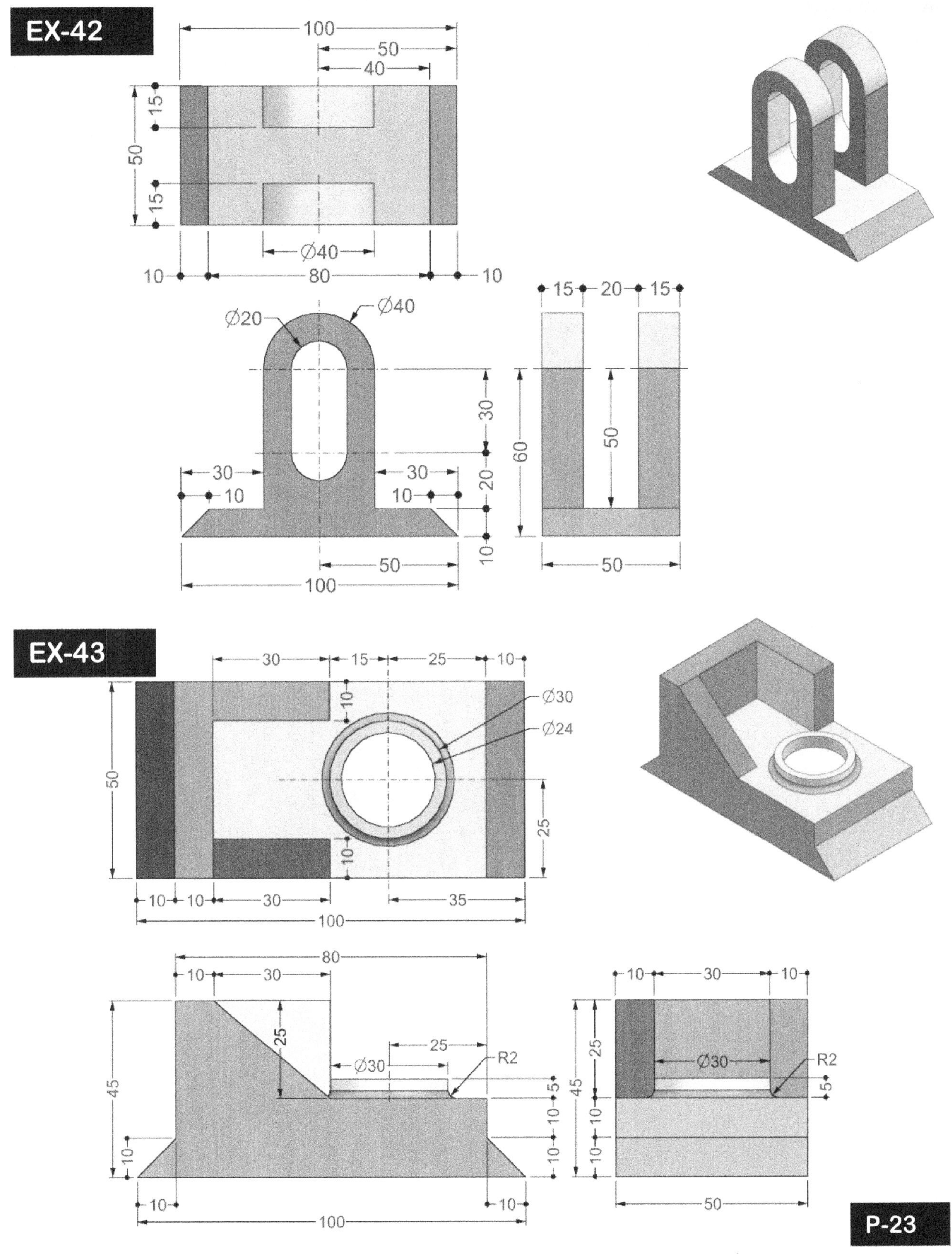

EX-42
100
50
40
15
50
15
Ø40
80
10
10
Ø20
Ø40
30
60
30
30
20
10
10
10
50
100
15
20
15
50
50

EX-43
30
15
25
10
10
Ø30
Ø24
50
25
10
10
30
35
100
80
10
30
25
25
Ø30
R2
45
10
Ø30
10
5
10
10
100
10
30
10
25
45
Ø30
R2
10
10
5
50

P-23

EX-44

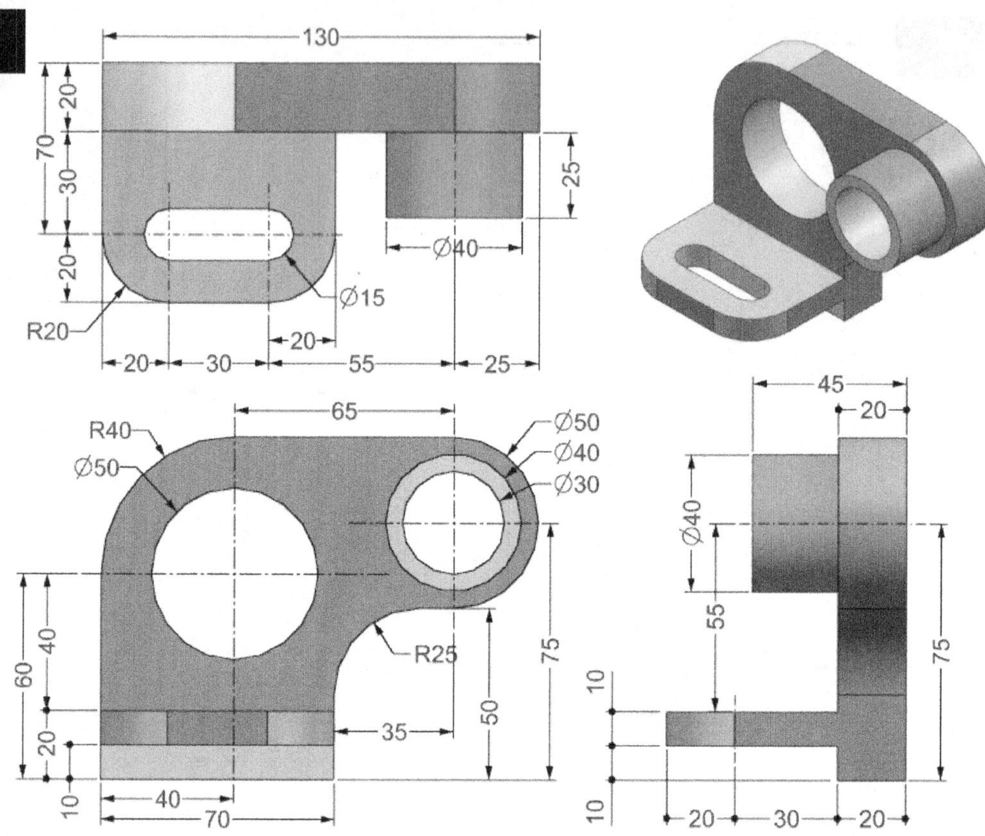
130
20
70
30
20
25
Ø40
R20
Ø15
20
30
20
55
25
R40
Ø50
Ø50
65
Ø50
Ø40
Ø30
60
40
R25
75
20
35
50
10
40
70
45
20
Ø40
55
75
10
10
20
30
20

EX-45

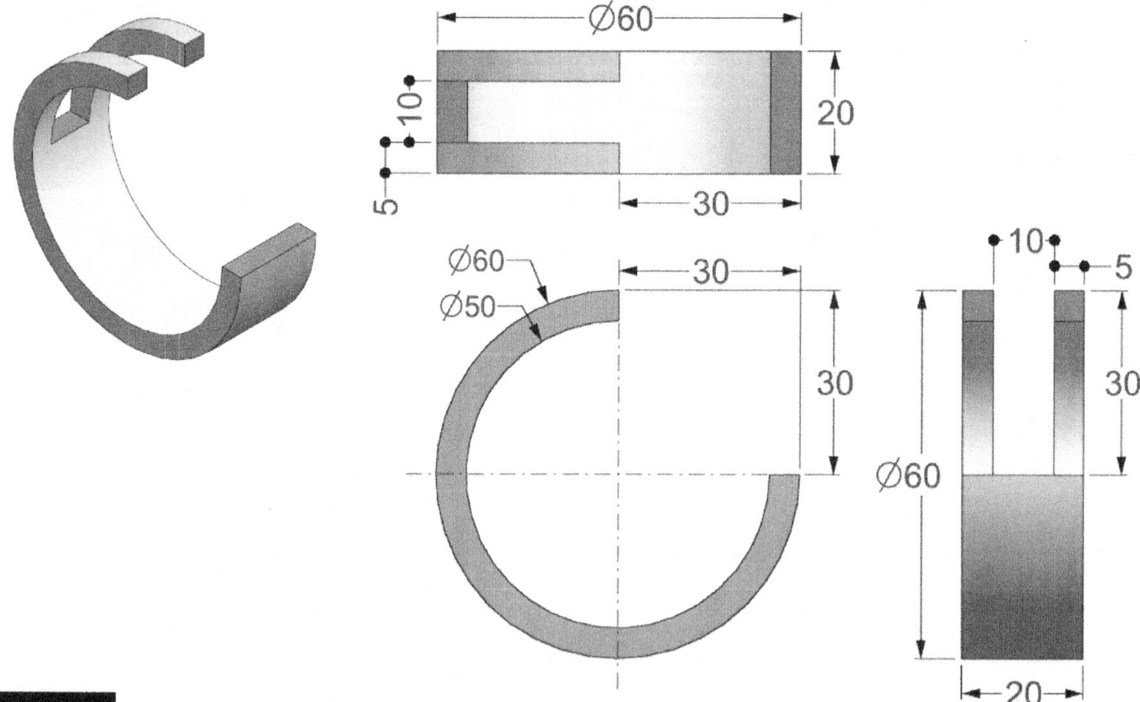
Ø60
10
20
5
30
Ø60
Ø50
30
30
30
10
5
30
Ø60
20

EX-46
2X Ø16
2X Ø12
2X R30
R15
20
80
30
120
30
60
35
60
45
30
20
50
30

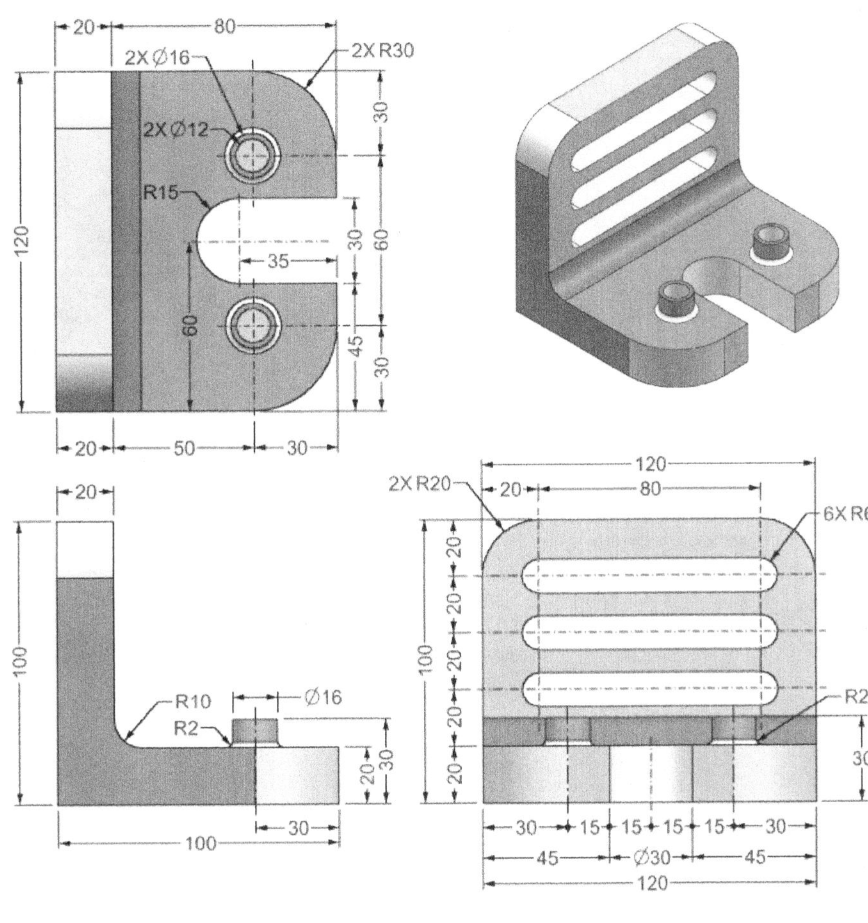

2X R20
20
120
80
6X R6
20
20
20
20
100
R2
30
30
15
15
15
15
30
45
Ø30
45
120

EX-47
20
100
R10
R2
Ø16
20
30
30
100

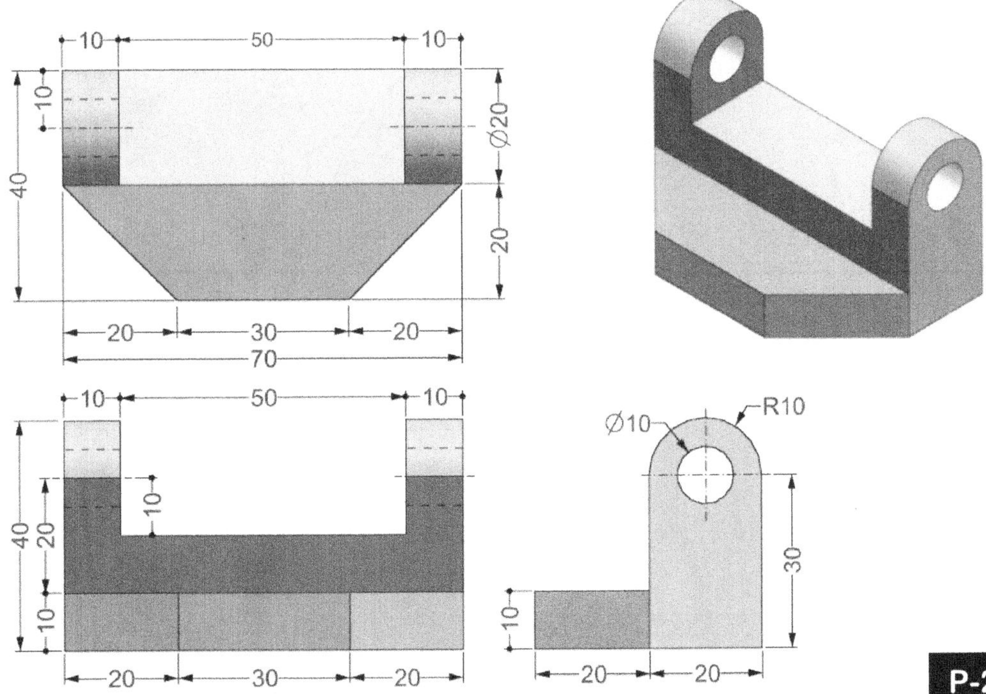

10
50
10
10
Ø20
40
20
20
30
20
70
10
50
10
10
40
20
10
20
30
20
Ø10
R10
30
10
20
20
P-25

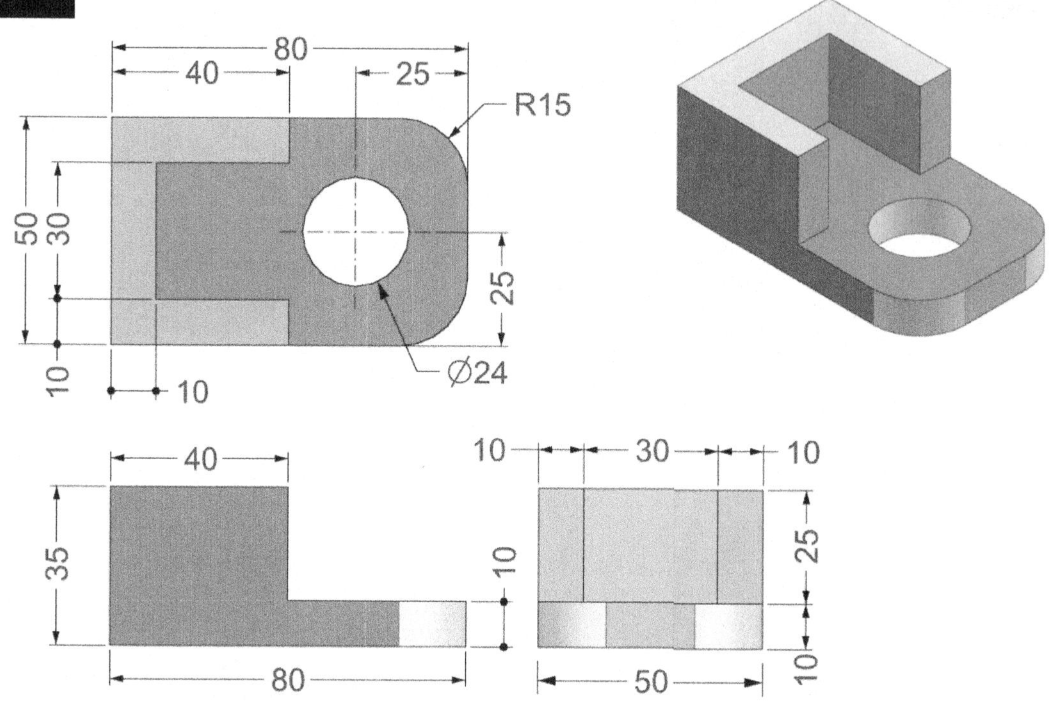

EX-48
80
40
25
R15
50
30
25
10
10
Ø24
40
35
80
10
30
10
10
25
50
10

EX-49

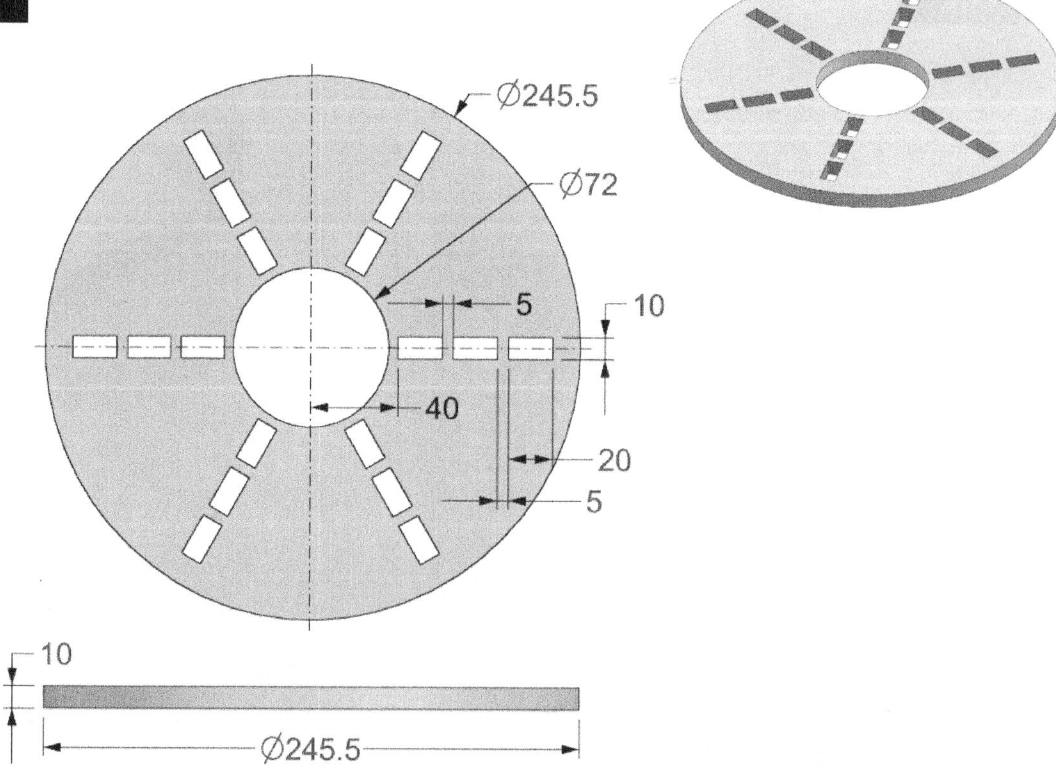

Ø245.5
Ø72
5
10
40
20
5
10
Ø245.5
P-26

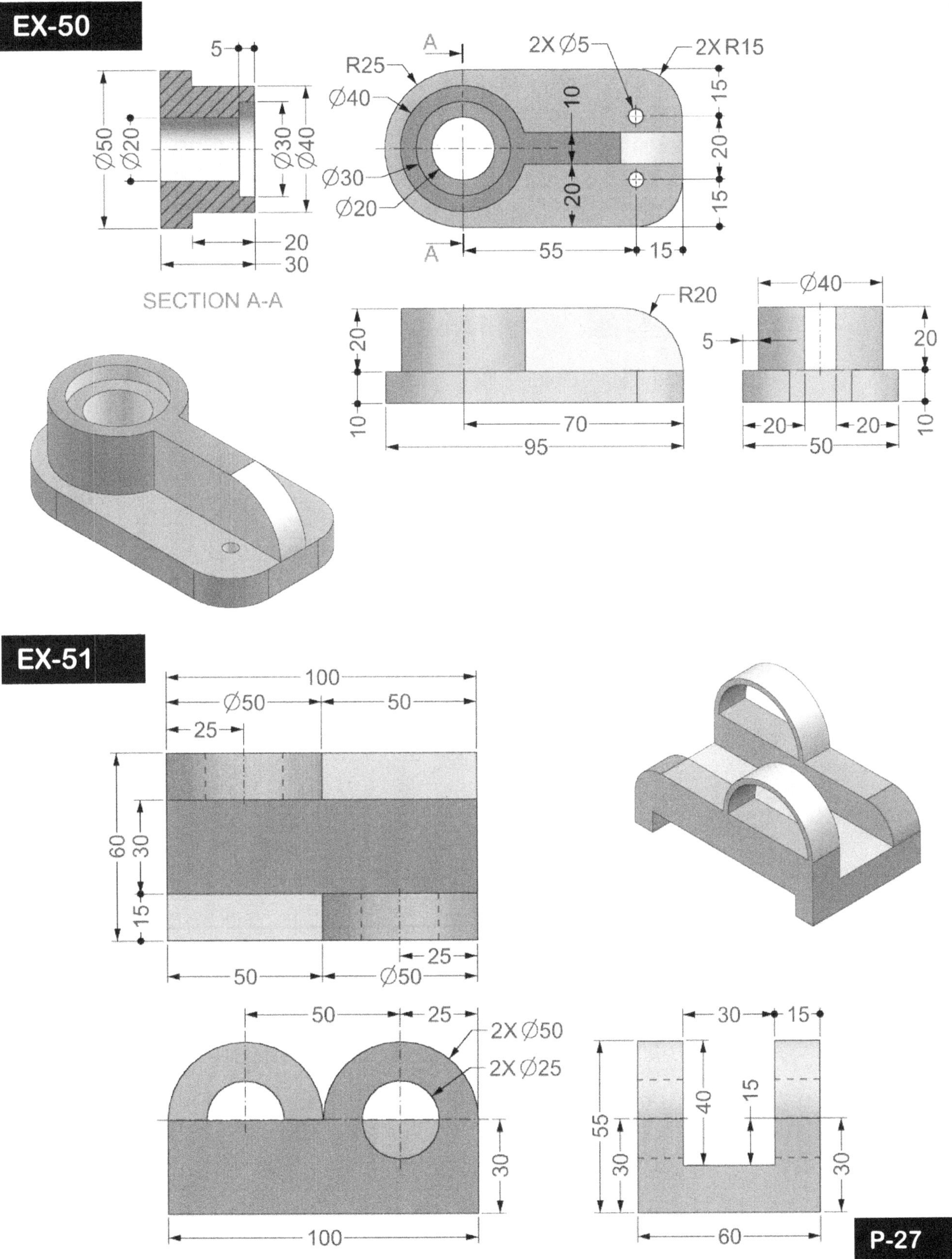
EX-50
5
Ø50
Ø20
Ø30
Ø40
20
30
SECTION A-A
R25
Ø40
Ø30
Ø20
A
A
2X Ø5
2X R15
10
15
20
15
20
55
15
R20
20
10
70
95
Ø40
5
20
20
50
20
10
EX-51
100
Ø50
50
25
60
30
15
50
25
Ø50
50
25
2X Ø50
2X Ø25
30
100
30
15
40
15
55
30
30
60
P-27

EX-52 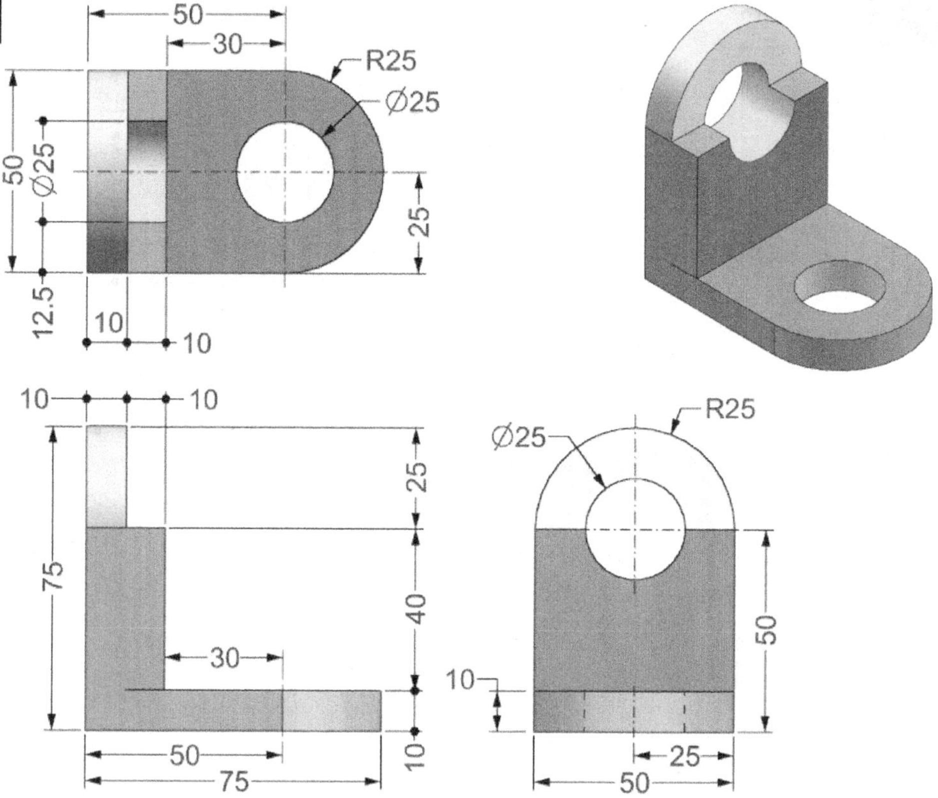
50
30
R25
Ø25
50
Ø25
25
12.5
10
10
10
10
25
75
40
30
50
75
10
Ø25
R25
50
10
25
50
EX-53 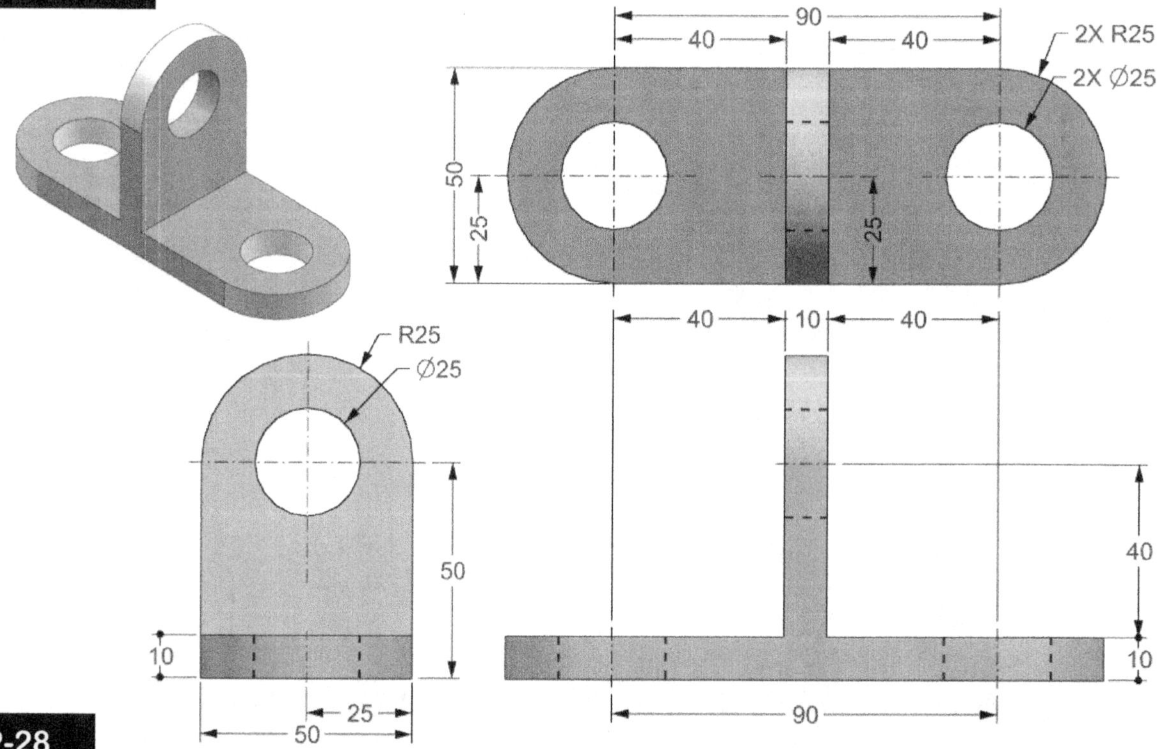
R25
Ø25
90
40
40
2X R25
2X Ø25
50
25
25
40
10
40
50
10
25
50
40
10
90
P-28

EX-54

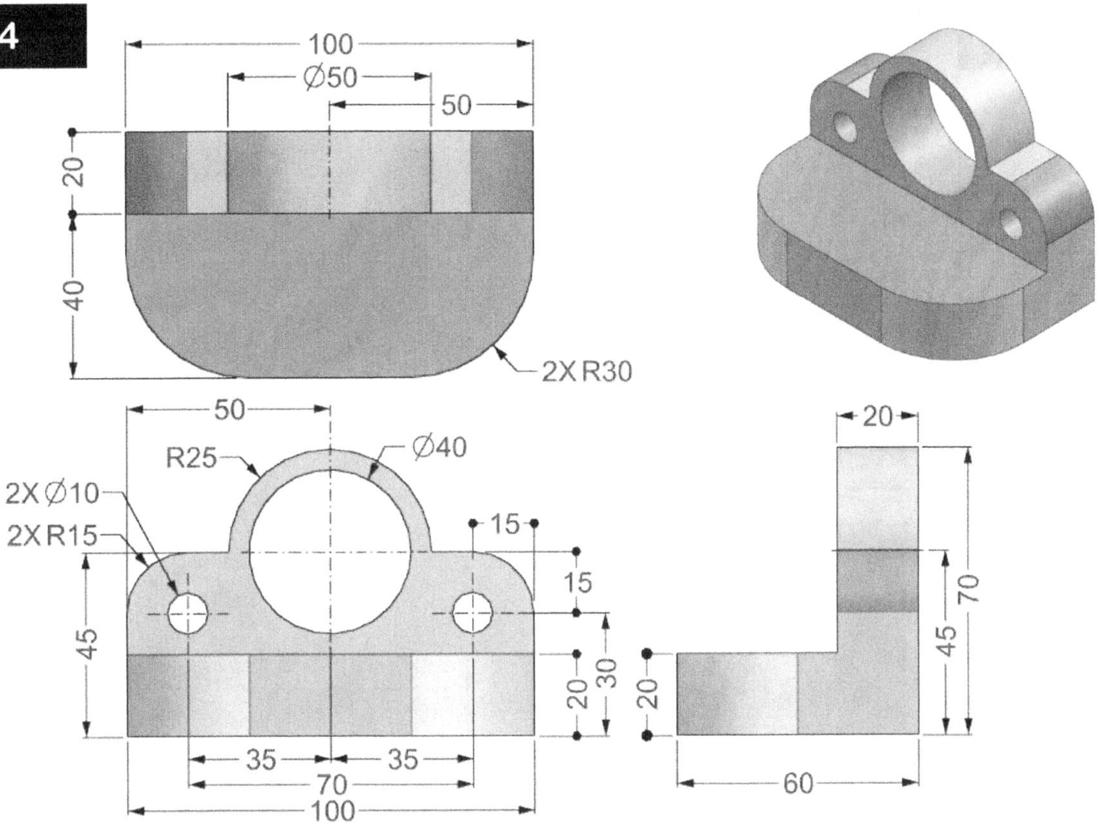

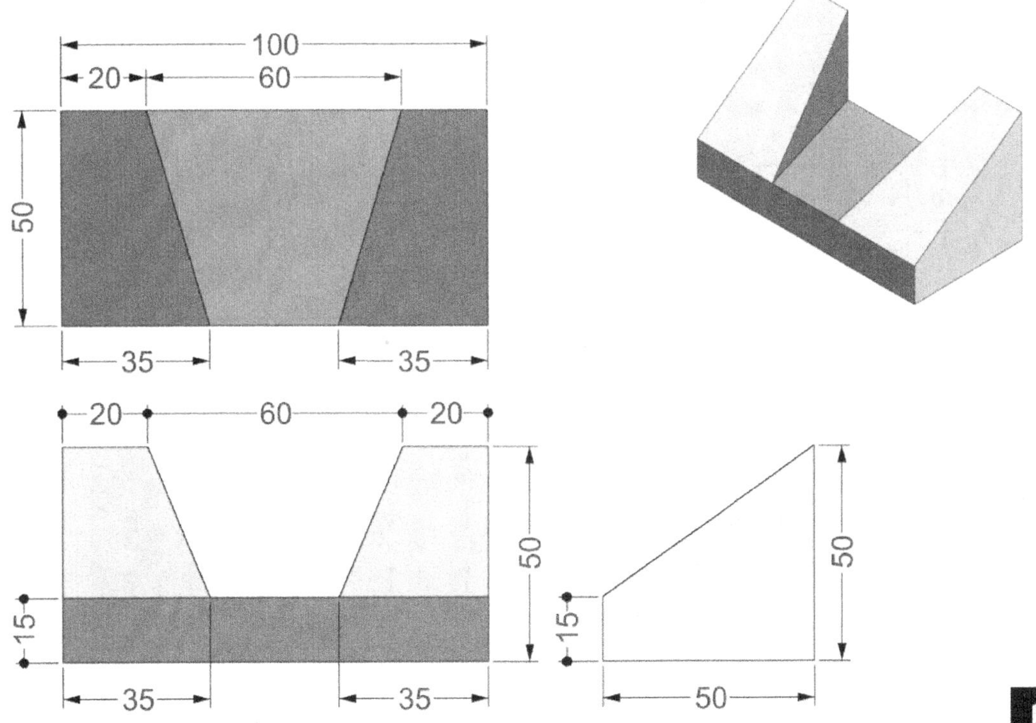

100
Ø50
50
20
40
2X R30
50
R25
Ø40
2X Ø10
2X R15
15
15
45
30
20
20
20
70
35
35
70
100
60
EX-55
100
20
60
50
35
35
20
60
20
50
50
15
35
35
15
50
50

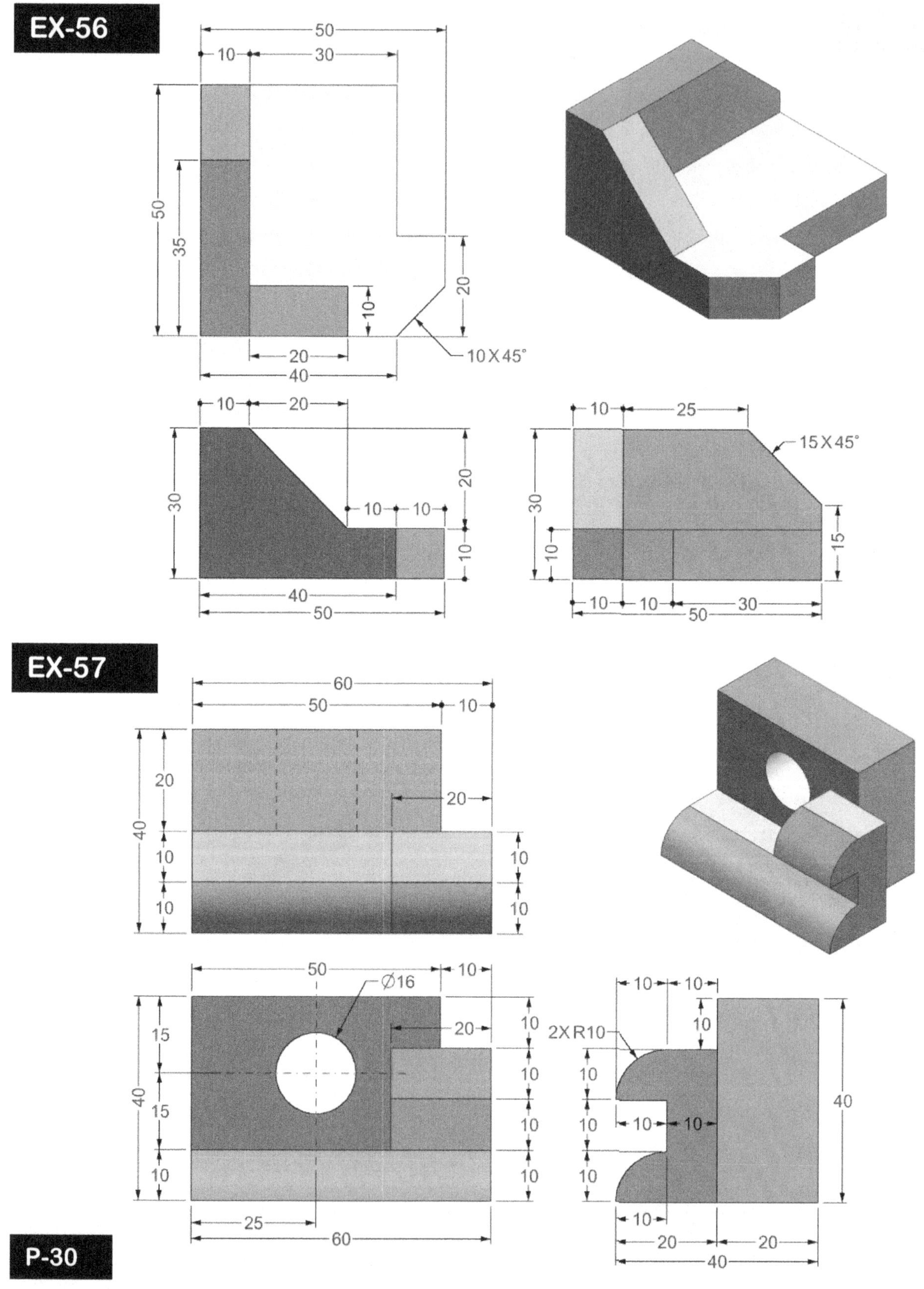

EX-56
50
10
30
50
35
20
10
20
40
10 X 45°
10
20
20
30
10
10
10
40
50
10
25
15 X 45°
30
10
15
10
10
50
30
EX-57
60
50
10
20
20
40
10
10
10
10
50
Ø16
10
20
10
15
10
40
10
15
10
25
60
10
10
10
2X R10
10
10
10
10
10
40
10
10
10
20
20
40
P-30

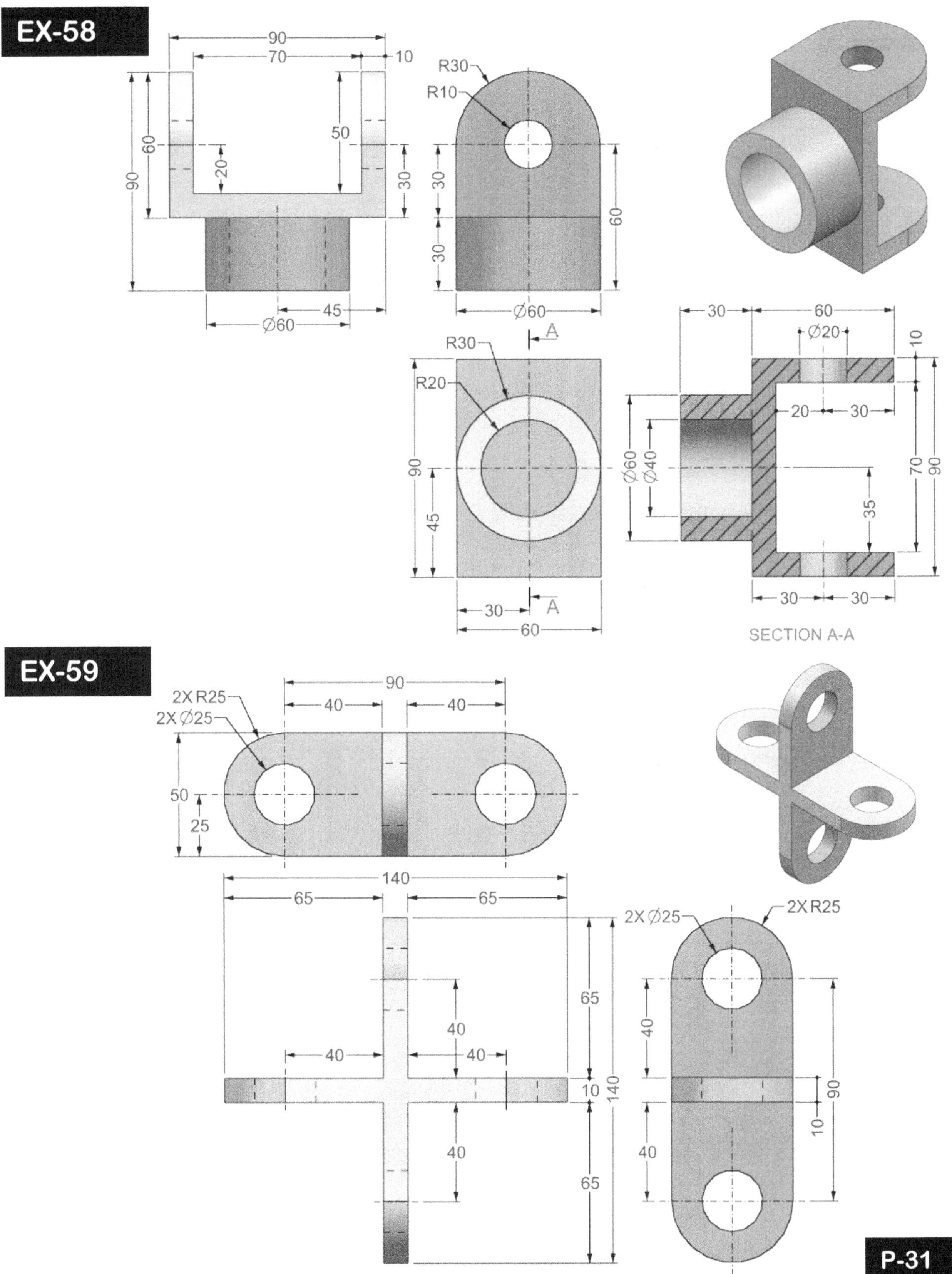

EX-58
90
70
10
60
50
90
20
30
45
Ø60
R30
R10
30
60
30
Ø60
A
R30
R20
90
45
30
60
A
SECTION A-A
30
60
Ø20
10
Ø60
Ø40
20
30
70
90
35
30
30
EX-59
2X R25
2X Ø25
90
40
40
50
25
140
65
65
65
40
40
40
40
65
10
140
2X Ø25
2X R25
40
90
10
40
P-31

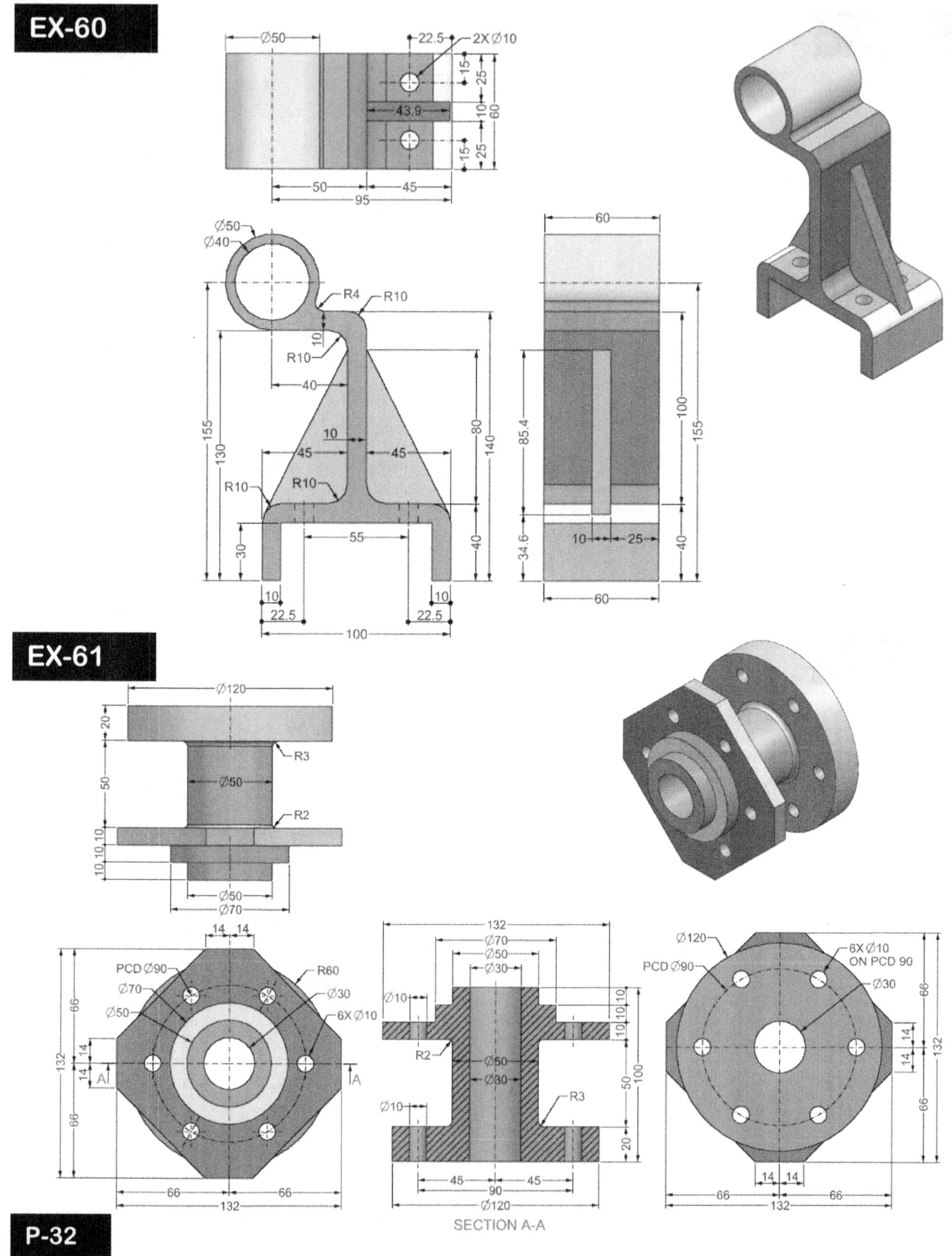

EX-60
Ø50
22.5
2X Ø10
15
25
60
43.9
10
15
25
50
45
95
Ø50
Ø40
R4
R10
R10
10
40
R10
10
45
45
155
130
R10
R10
80
140
55
30
40
10
22.5
10
22.5
100
60
85.4
100
155
34.6
10
25
40
60
EX-61
Ø120
20
R3
50
Ø50
R2
10 10 10
Ø50
Ø70
14 14
PCD Ø90
Ø70
Ø50
R60
Ø30
6X Ø10
66
132
14
14
14
A
66
66
66
132
132
Ø70
Ø50
Ø30
Ø10
10 10 10
R2
Ø50
Ø30
50
100
Ø10
R3
45 45
90
20
Ø120
SECTION A-A
Ø120
PCD Ø90
6X Ø10
ON PCD 90
Ø30
66
66
14
14
132
14 14
66
66
132
P-32

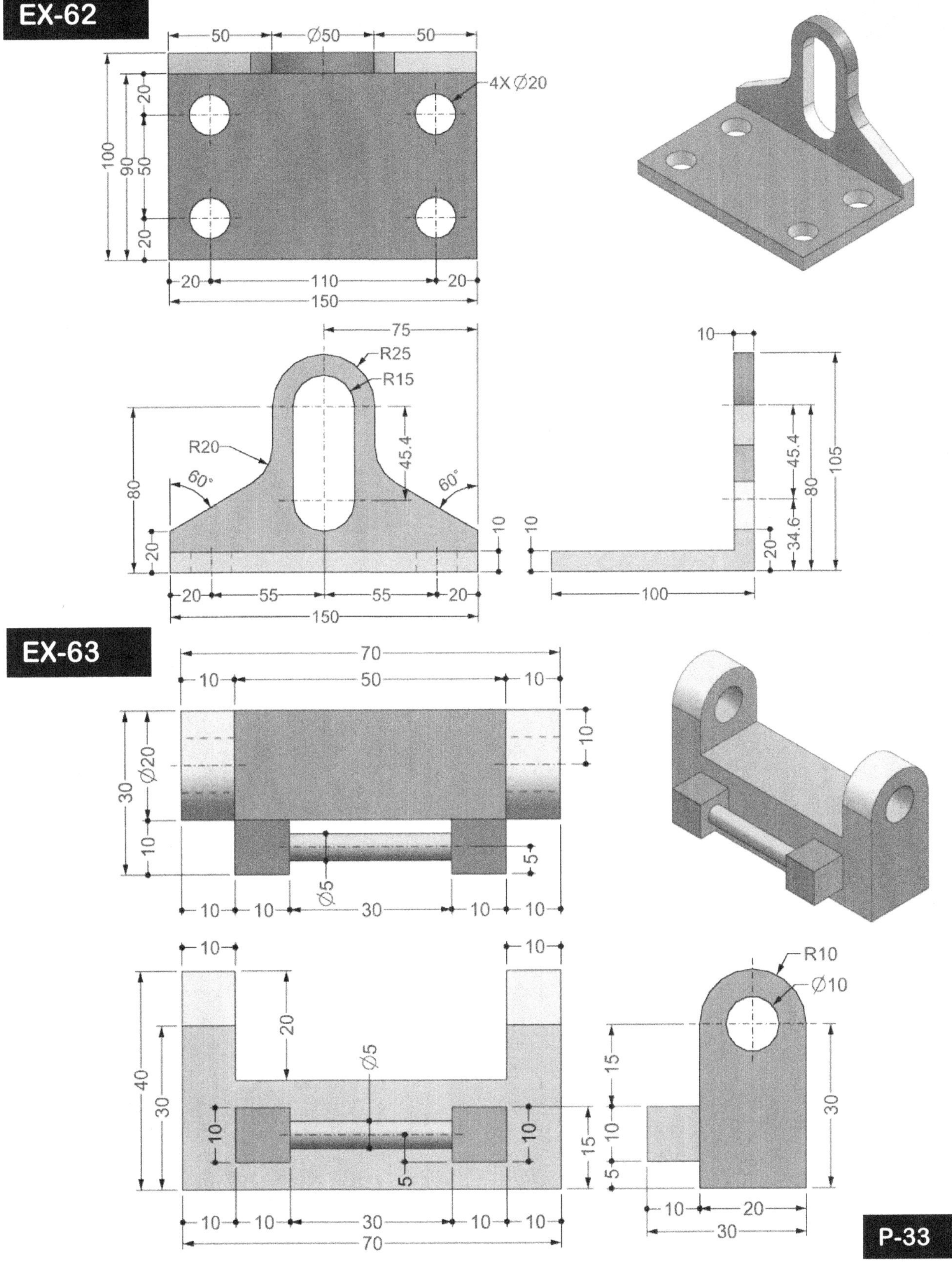

EX-62
50
Ø50
50
4X Ø20
20
100
90
50
20
20
110
20
150
75
R25
R15
R20
45.4
80
60°
60°
20
10
20
55
55
20
150
10
10
10
45.4
80
105
20
34.6
100
EX-63
70
10
50
10
10
30
Ø20
10
Ø5
5
10
10
30
10
10
10
10
20
Ø5
40
30
10
10
5
15
R10
Ø10
15
10
5
10
10
30
10
10
70
10
20
30
P-33

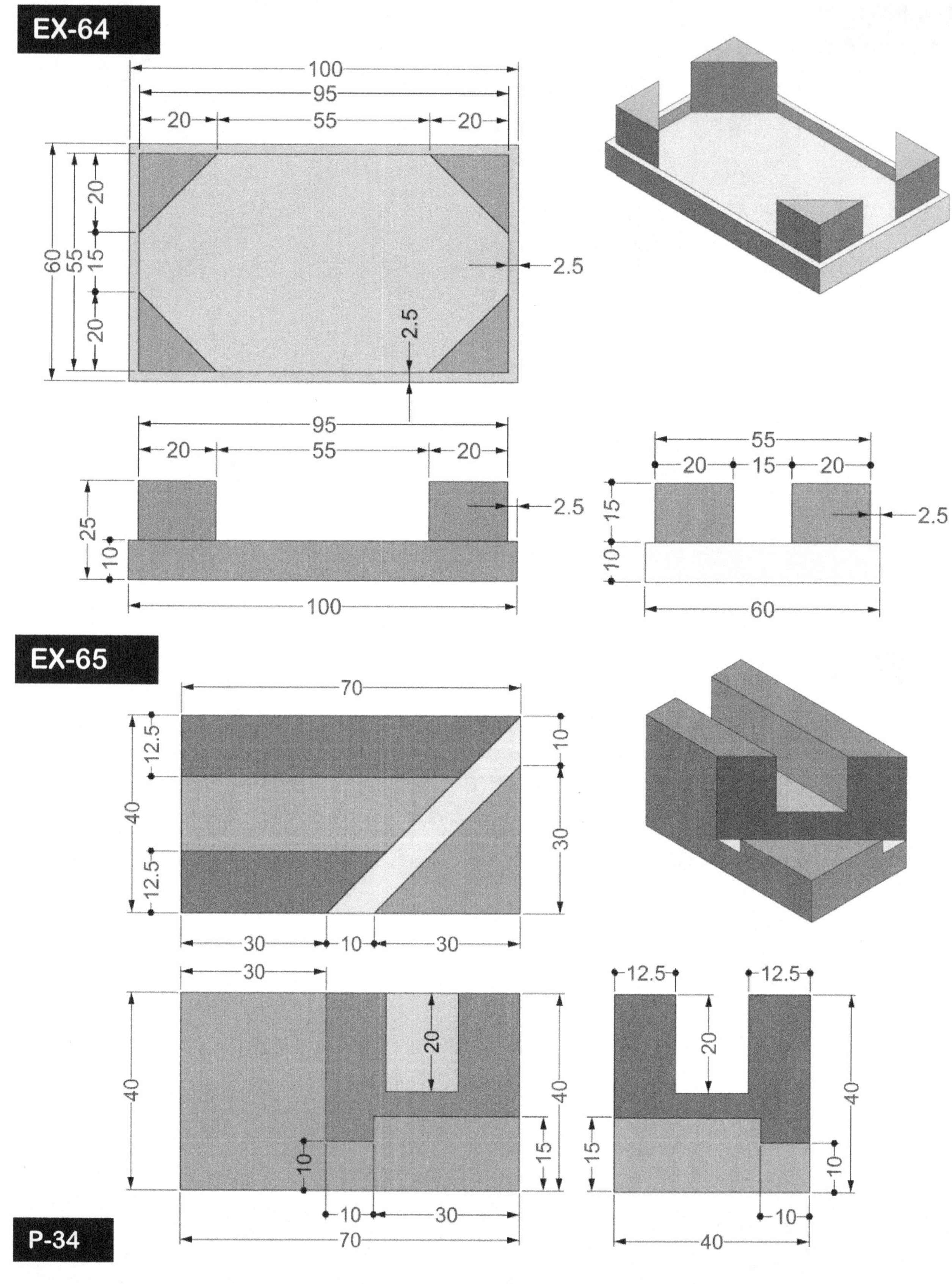

EX-64
100
95
20
55
20
20
55
15
60
20
2.5
2.5
95
20
55
20
25
10
2.5
55
20
15
20
15
10
2.5
100
60
EX-65
70
12.5
10
40
30
12.5
30
10
30
30
20
40
40
15
10
10
30
70
12.5
12.5
20
40
15
15
10
10
40
P-34

EX-66
30
R20
R20
R10
40
20
30
10
10
50
R10
10
R20
50
20
40
10
30
10
40
20

EX-67
70
45
R20
Ø20
10
10
15
20
40
10
45
65
45
Ø20
R25
40
30
10
30
20
10
Ø40
10
25
30
50
10
20

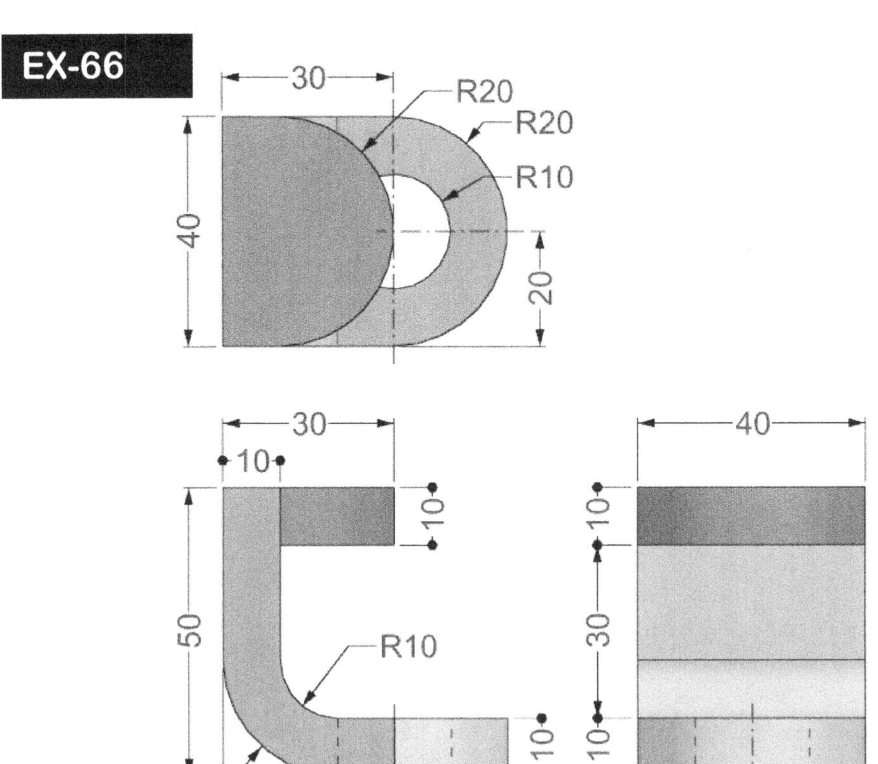

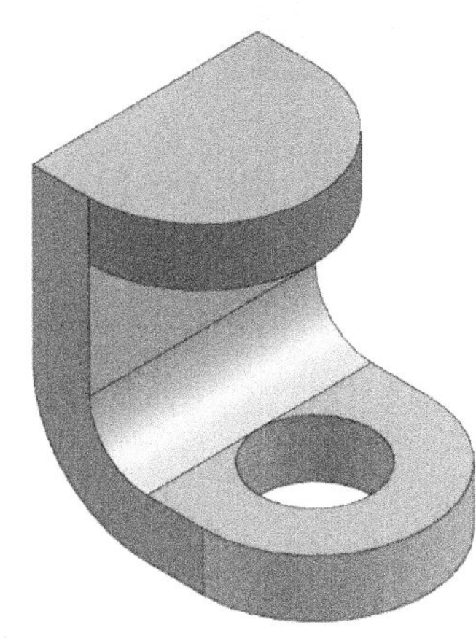

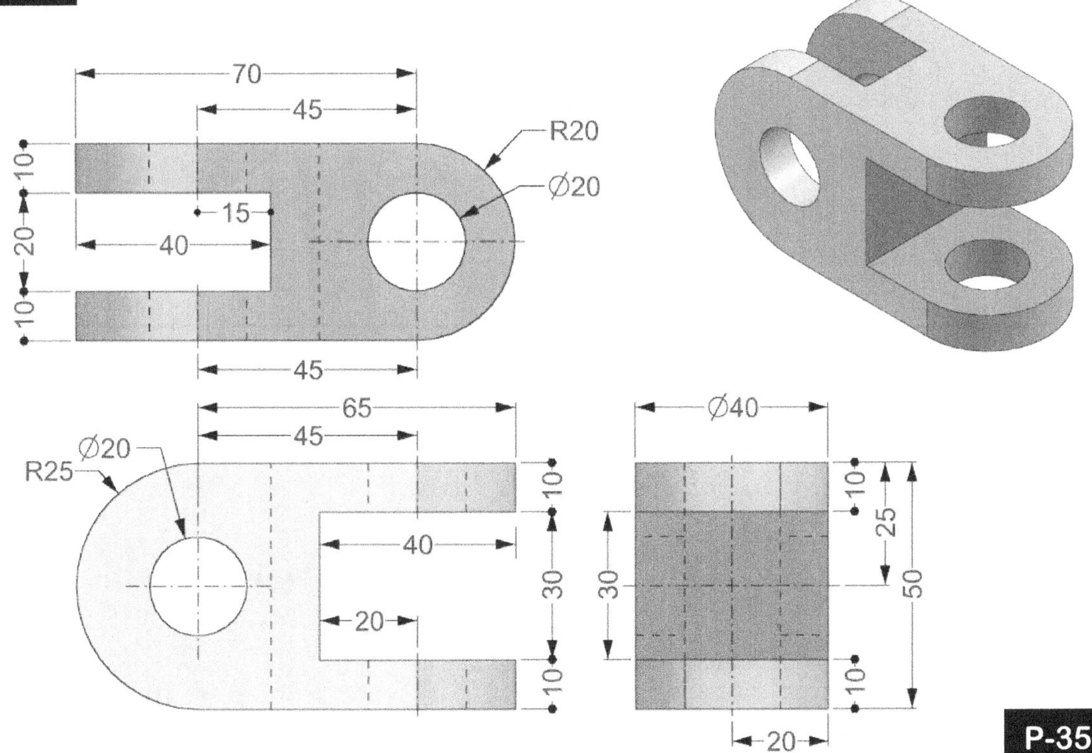

P-35

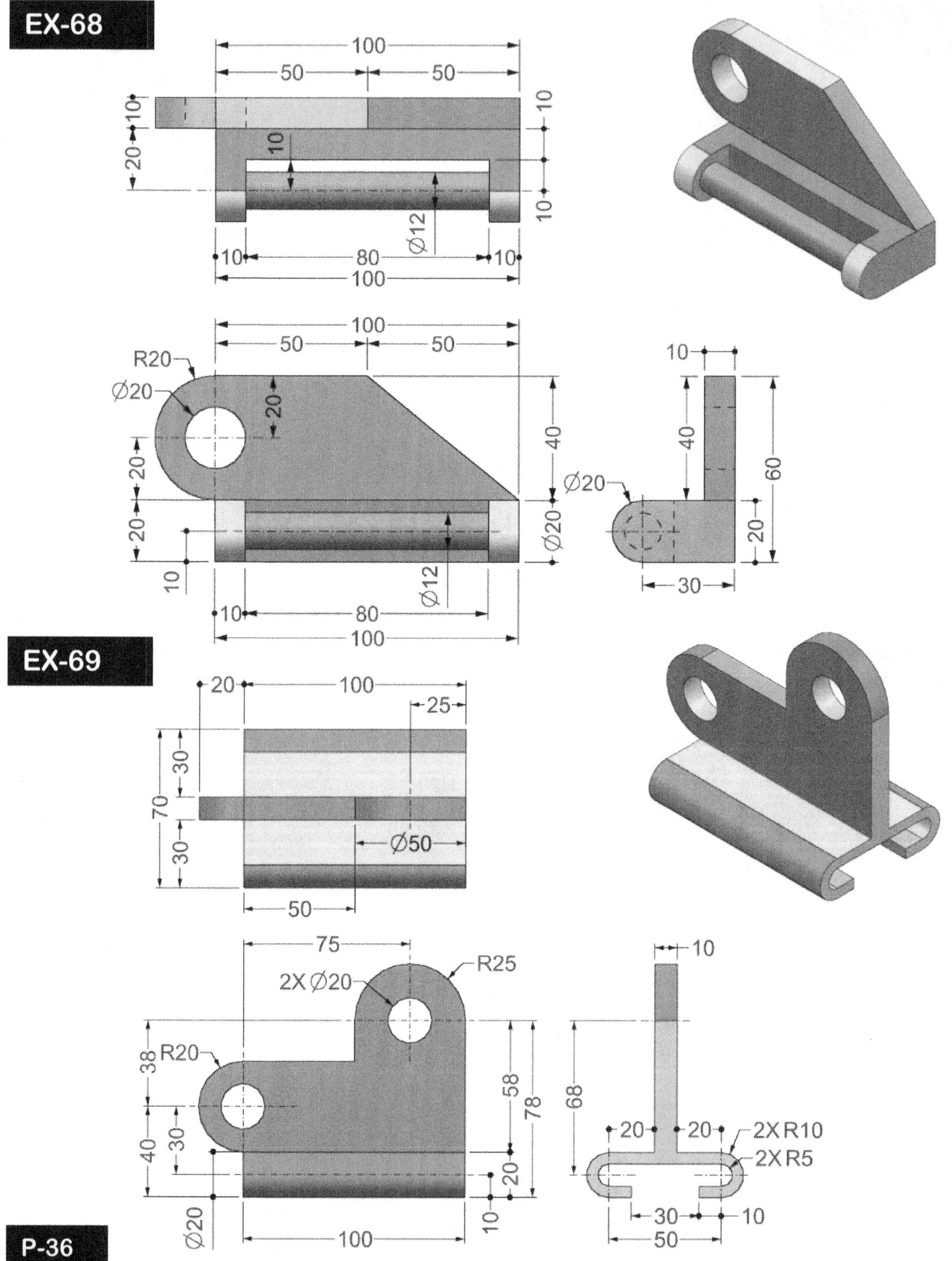

EX-68
EX-69
P-36
100
50
50
10
20
10
10
10
80
10
100
Ø12
R20
Ø20
100
50
50
20
20
20
20
40
Ø20
Ø20
Ø12
10
10
80
100
10
40
60
20
Ø20
30
20
100
25
30
70
30
Ø50
50
75
2X Ø20
R25
R20
38
58
78
40
30
20
Ø20
100
10
68
10
20
20
2X R10
2X R5
30
10
50

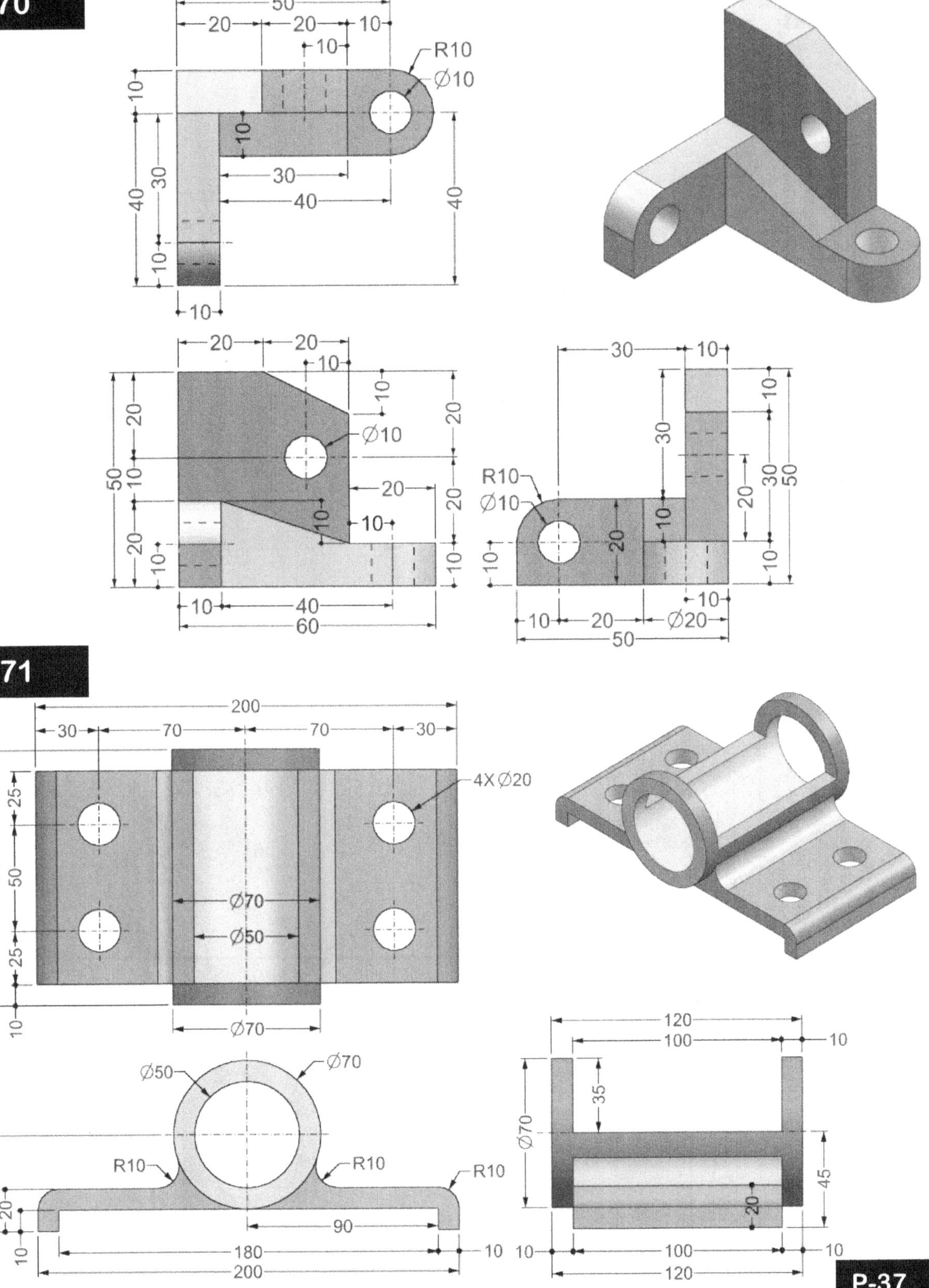

50
20
20
10
10
R10
Ø10
10
10
10
30
40
30
40
40
10
20
20
10
20
Ø10
20
50
10
20
20
10
10
10
10
10
40
10
60
30
10
30
10
30
30
50
R10
Ø10
20
10
20
10
10
10
10
20
Ø20
50

200
30
70
70
30
4X Ø20
25
120
50
Ø70
Ø50
25
10
Ø70
Ø50
Ø70
R10
R10
R10
45
20
10
90
180
10
200
120
100
10
35
Ø70
45
20
10
100
10
10
120

EX-72
25
13
15
140
80
10
27
6
27
28
5
25
13
15
10
10
80
EX-73
15 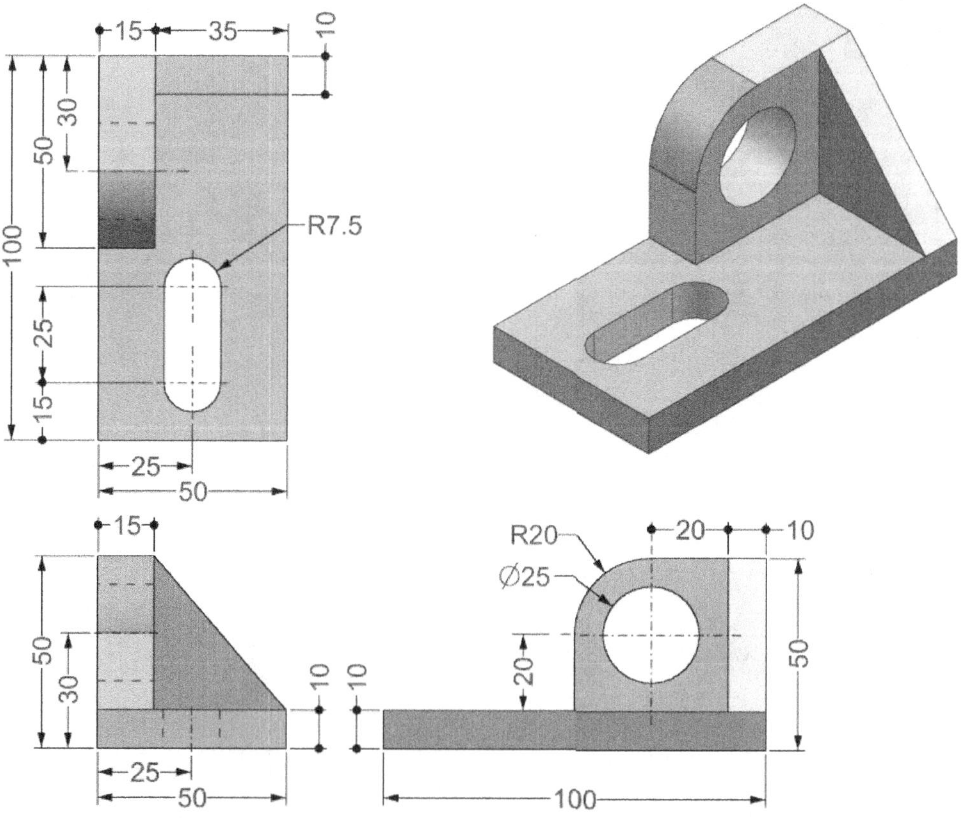
35
10
30
50
100
R7.5
25
15
25
50
15
50
30
10
25
50
R20
Ø25
20
10
20
50
100
10
10
P-38

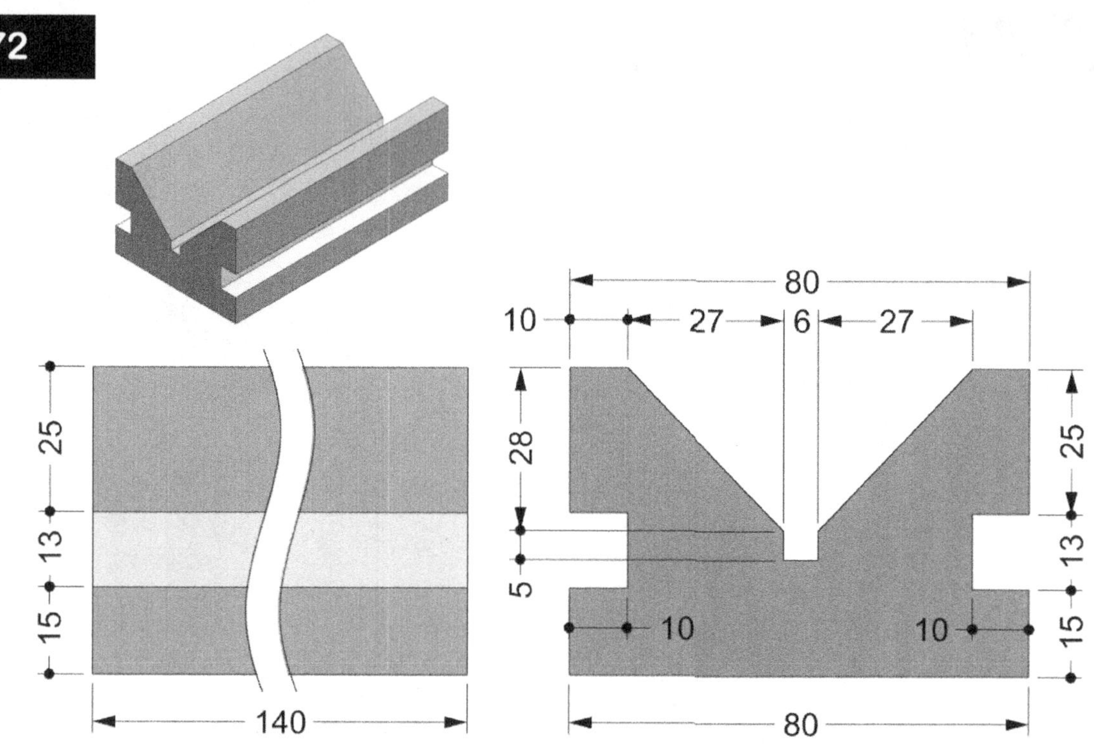

EX-74

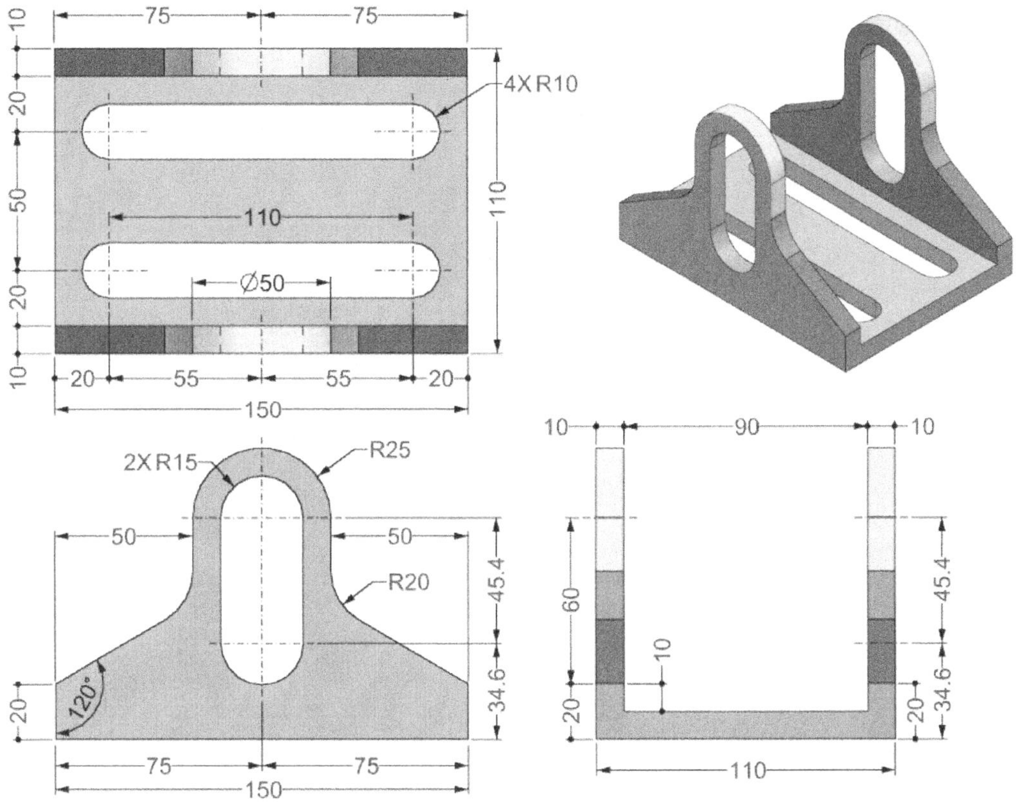

10
75
75
4X R10
20
50
110
110
20
10
20
55
55
20
150
2X R15
R25
50
50
R20
45.4
20
120°
34.6
75
75
150
10
90
10
60
10
20
45.4
20
34.6
110

EX-75

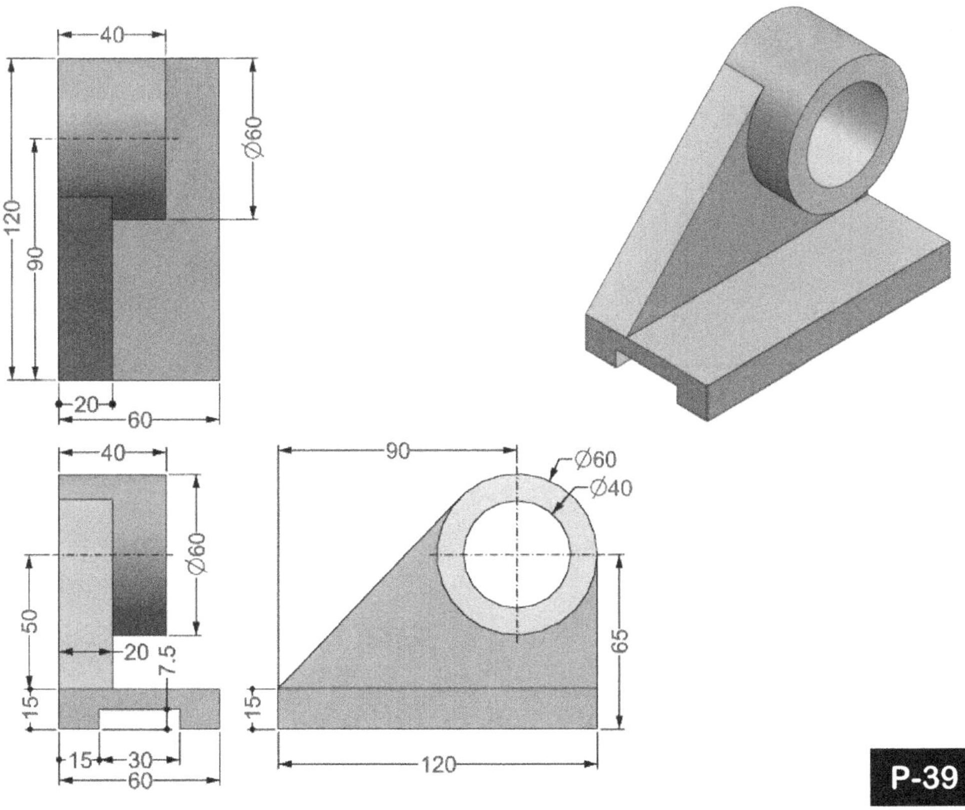

40
Ø60
120
90
20
60
40
Ø60
50
20
7.5
15
15
30
60
Ø60
Ø40
90
65
15
120

P-39

EX-76 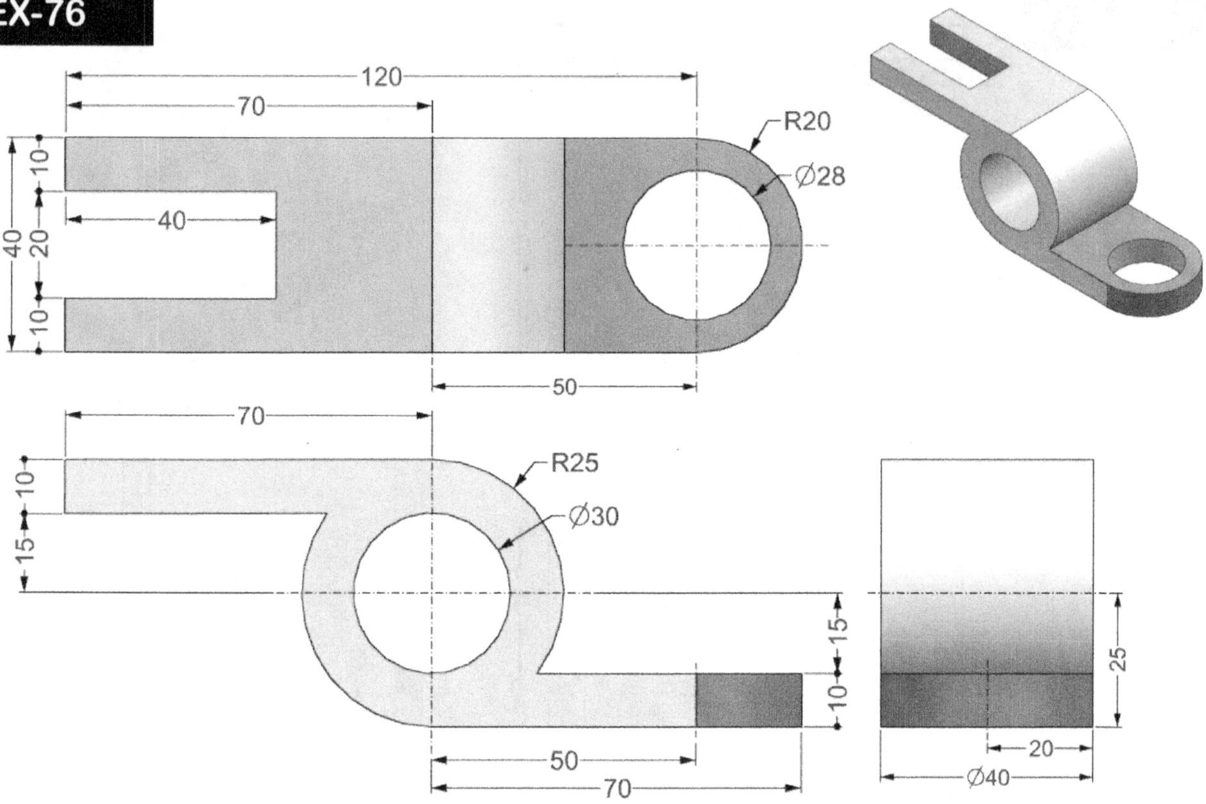
120
70
10
40
20
10
40
R20
Ø28
50
70
10
15
R25
Ø30
15
50
10
70
25
20
Ø40
EX-77
R15
R25
50
R10
R5
A
A
50
Ø30
R1
30
30
R4
20
30
R2
Ø10
5
SECTION A-A
P-40

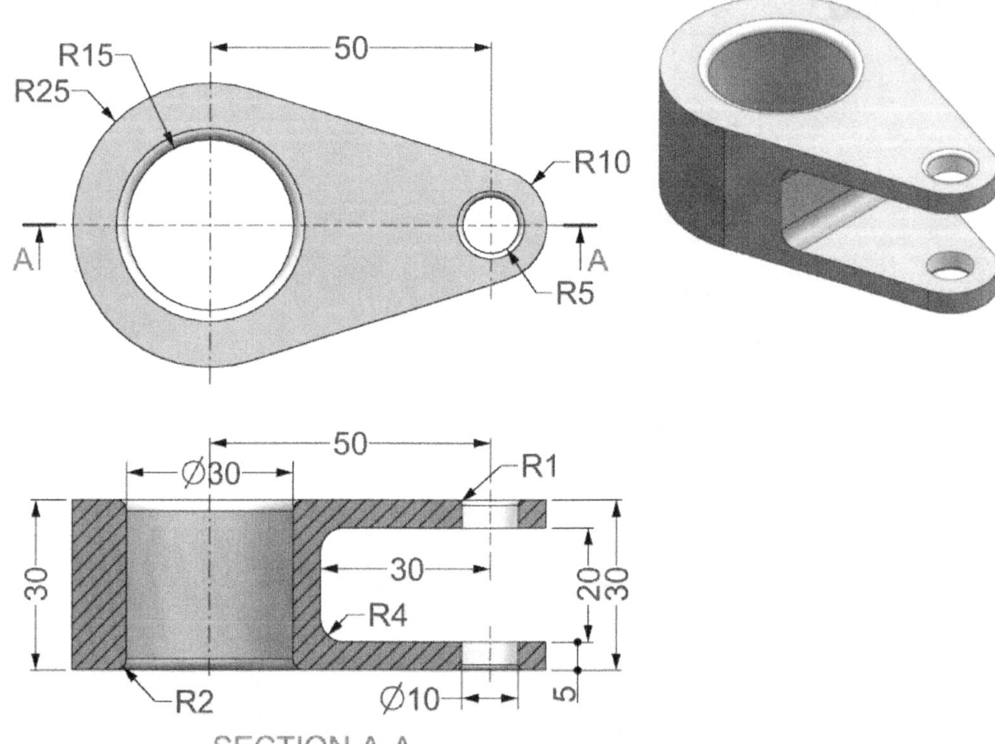

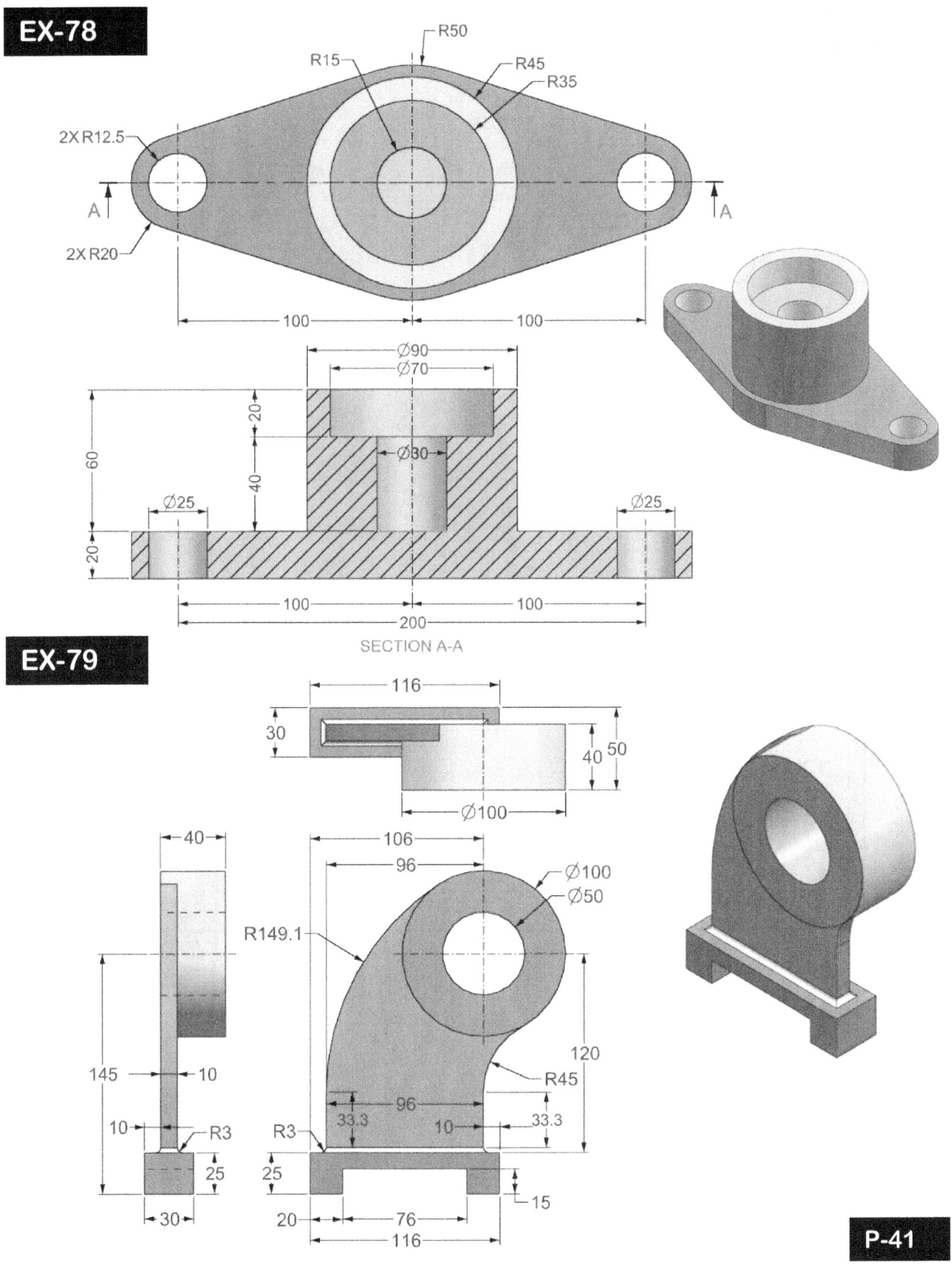
EX-78
R50
R15
R45
R35
2X R12.5
2X R20
A
A
100
100
Ø90
Ø70
Ø30
20
40
60
Ø25
Ø25
20
100
100
200
SECTION A-A
EX-79
116
30
40
50
Ø100
40
106
96
Ø100
Ø50
R149.1
145
10
120
R45
96
33.3
10
33.3
R3
R3
25
25
15
30
20
76
116
P-41

EX-80
6 HOLES,⌀10
ON DIA 32 PCD
4 HOLES,⌀8.6
ON DIA 54 PCD
⌀70
⌀16
A
A
⌀54
⌀32

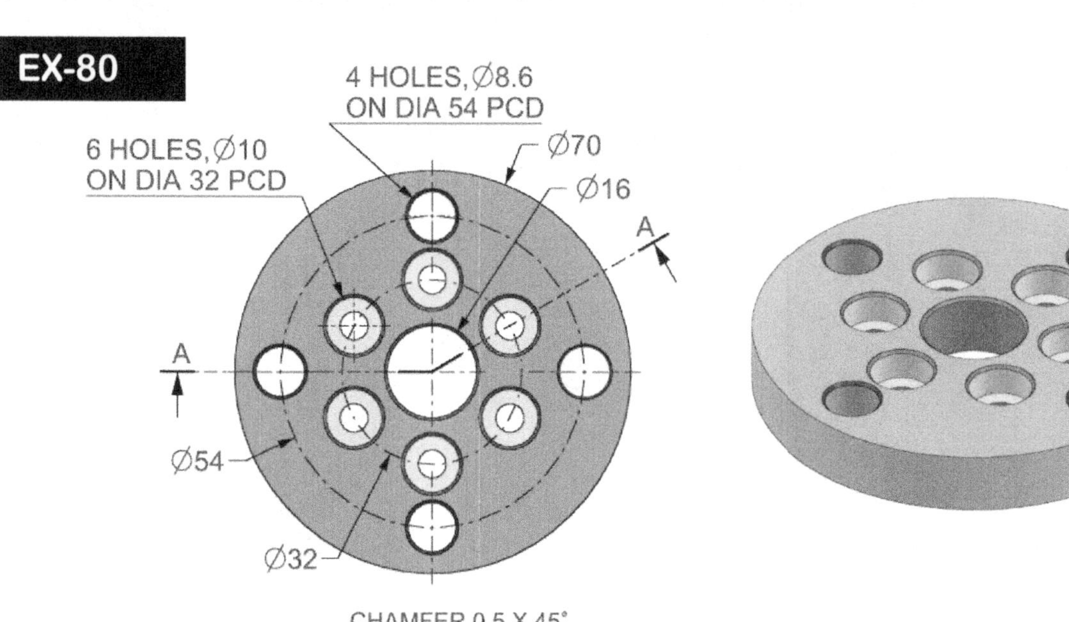

CHAMFER 0.5 X 45°
4X ⌀8.6
⌀16
6X ⌀10
10
5
5
SECTION A-A
(SCALE 1:1)

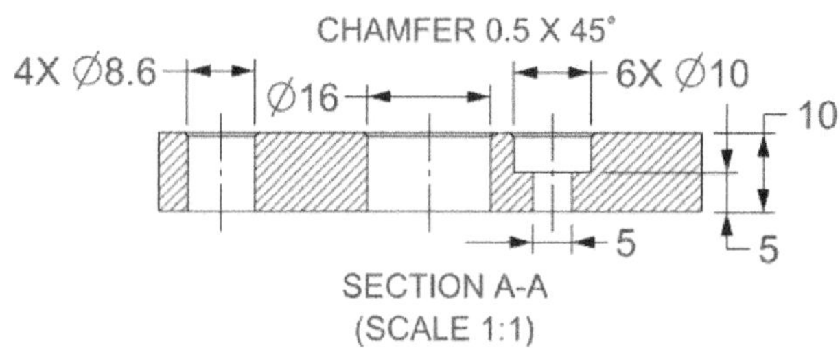

EX-81
10 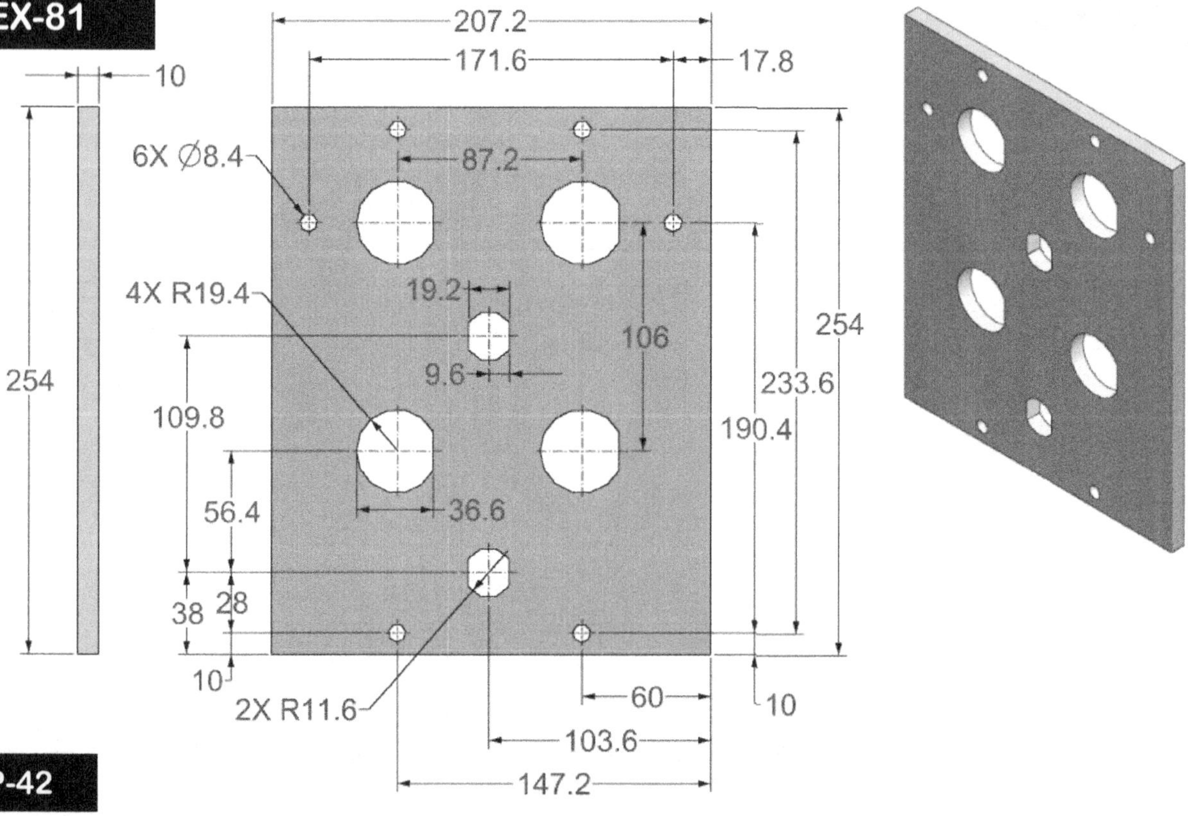
6X ⌀8.4
4X R19.4
109.8
56.4
38 28
10
2X R11.6
207.2
171.6
17.8
87.2
19.2
9.6
36.6
106
254
233.6
190.4
254
60
10
103.6
147.2

P-42

EX-82
1 X 45°
R3
1 X 45°
Ø40
Ø20
20
30
Ø40
Ø9
A
A
A/F 20
1x45°
Ø40
10
20
0.6X45°
R3
30
0.6X45°
Ø9
Ø20
SECTION A-A
(SCALE 1:1)

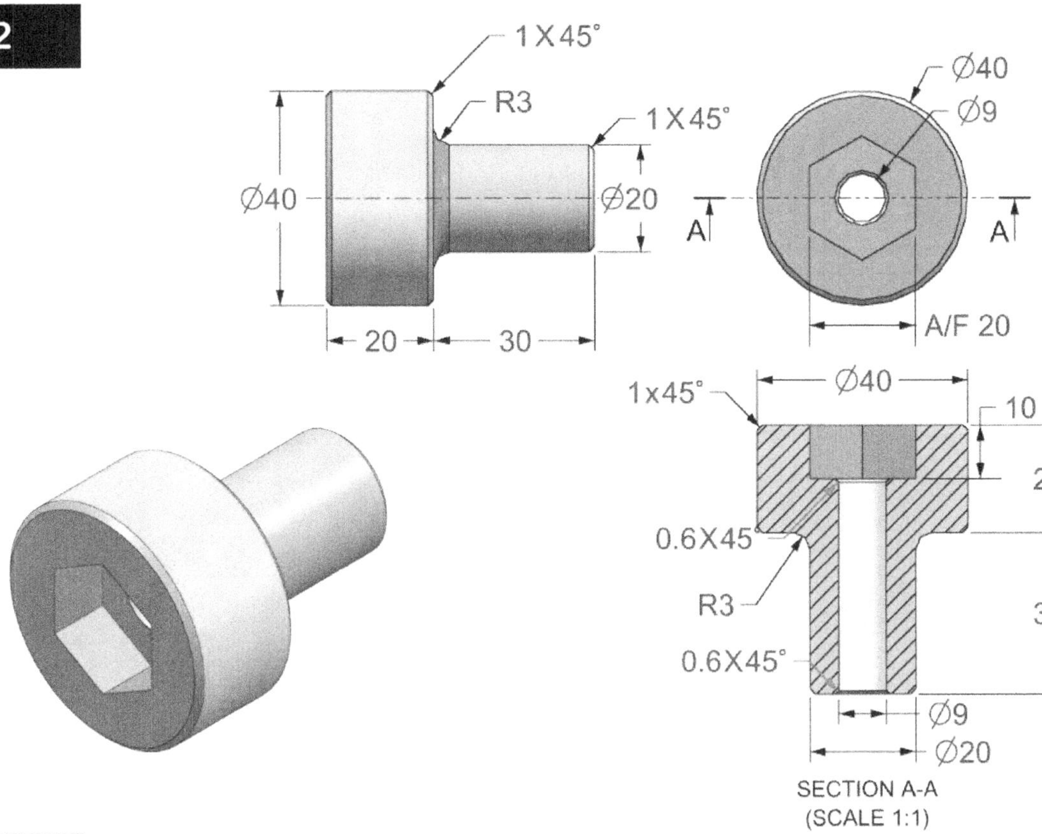

EX-83
A
30
65
R20
2X Ø40
2X Ø30
Ø30
Ø40
10
45
10
100
Ø12
Ø20
Ø55
R98
Ø55
4X Ø12
Ø12
30
PCD Ø38
A
SECTION A-A
(SCALE 1:1)

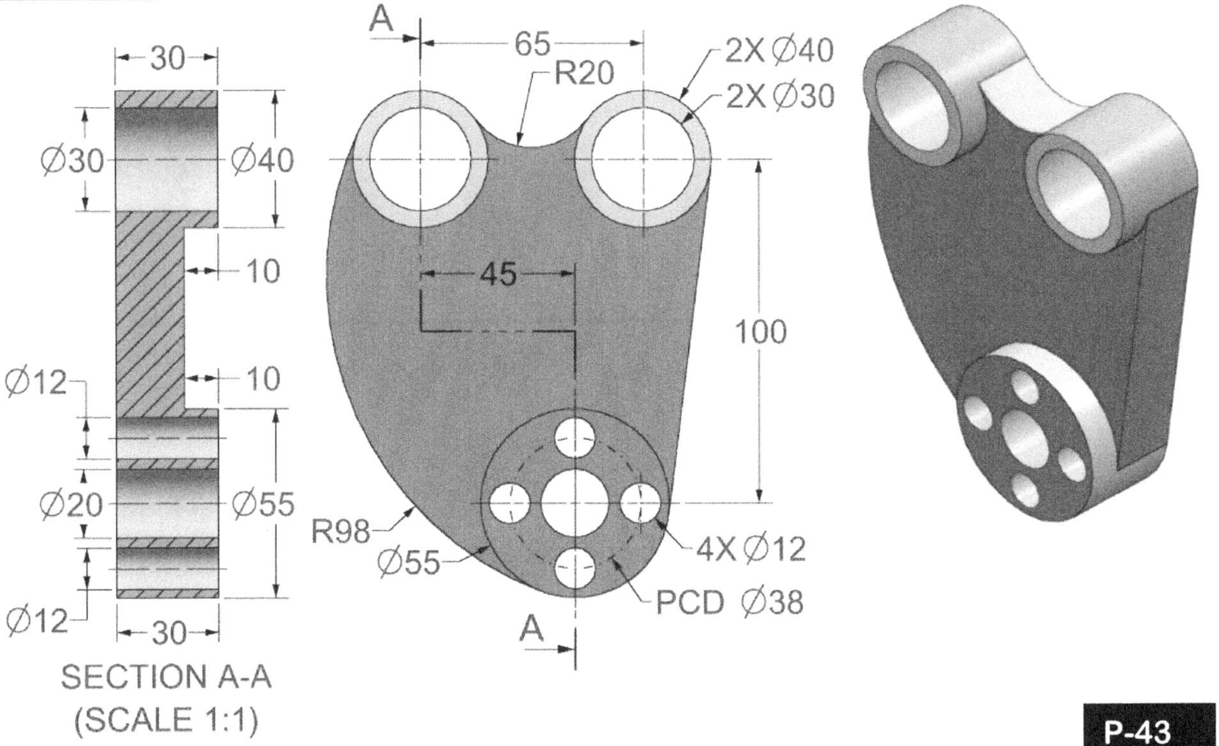

EX-84

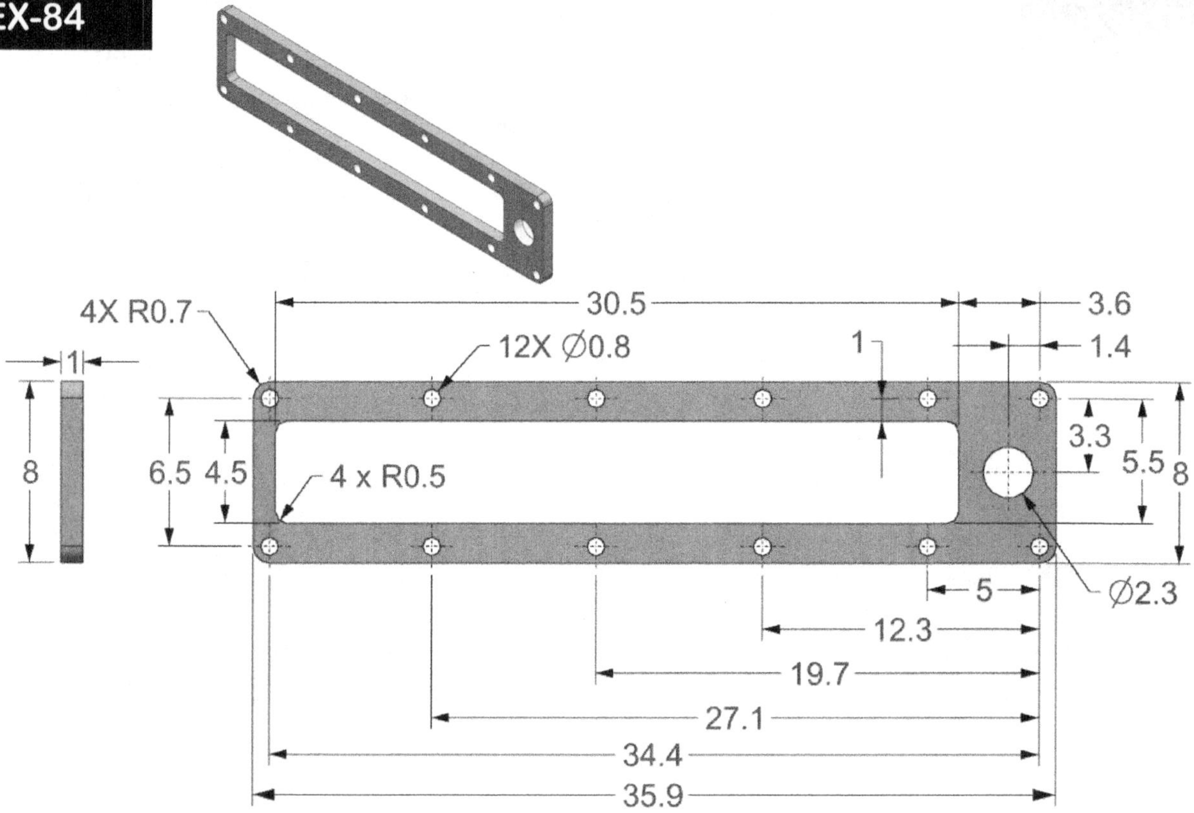

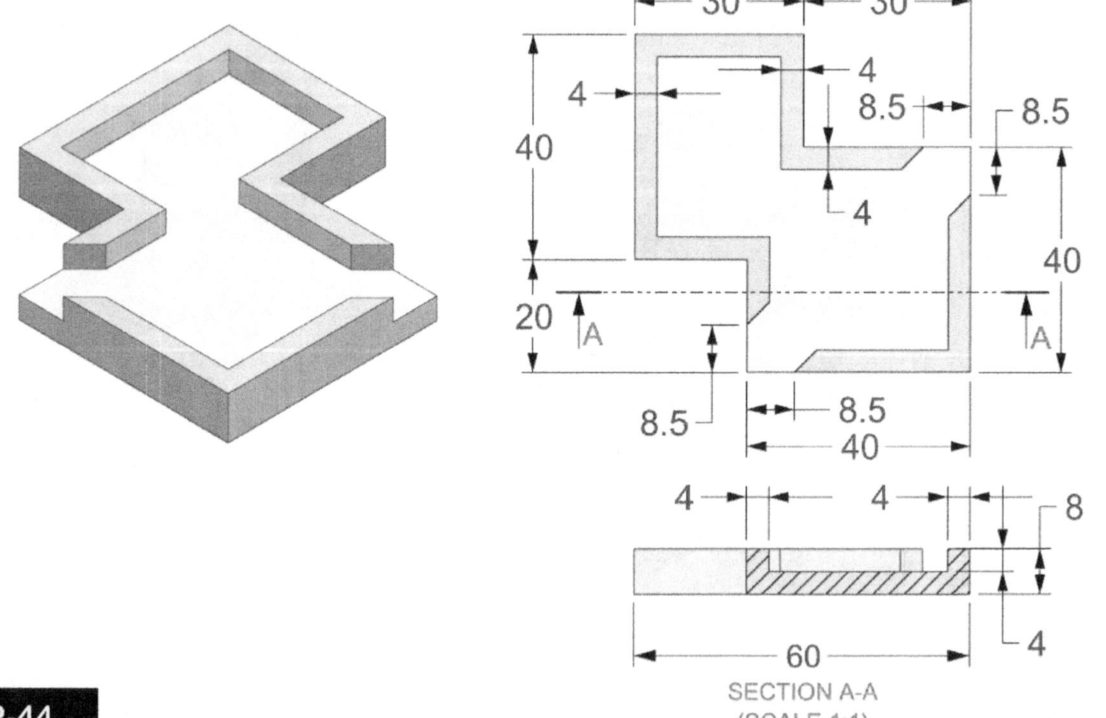
EX-84
4X R0.7
12X Ø0.8
1
3.6
1.4
1
3.3
5.5
8
6.5
4.5
8
4 x R0.5
Ø2.3
5
12.3
19.7
27.1
34.4
35.9
30.5
EX-85
30
30
4
4
8.5
8.5
40
4
40
20
A
A
8.5
8.5
40
4
4
8
60
4
SECTION A-A
(SCALE 1:1)
P-44

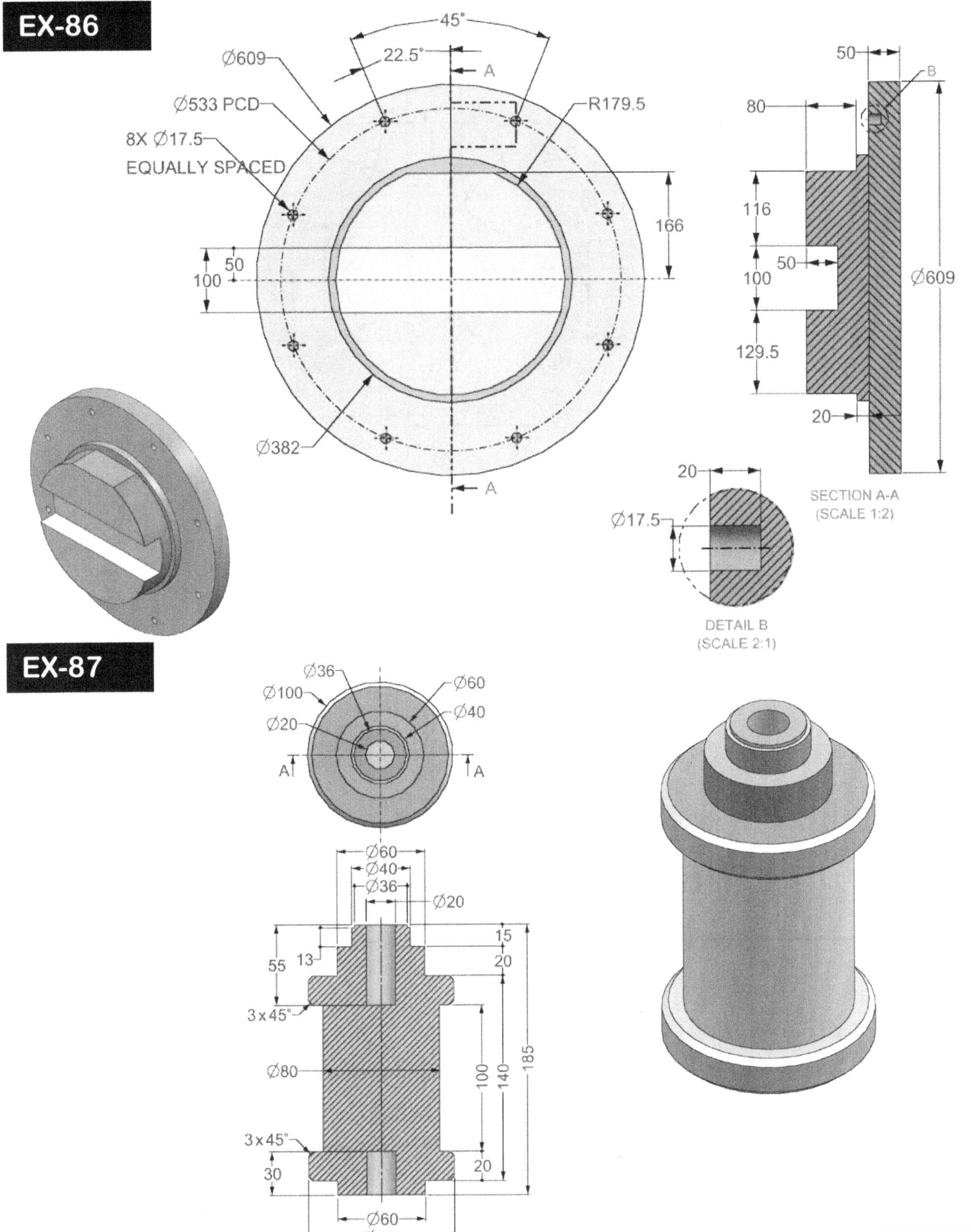

EX-86
EX-87
P-45
45°
22.5°
Ø609
Ø533 PCD
8X Ø17.5
EQUALLY SPACED
A
R179.5
166
100
50
Ø382
A
50
80
116
50
100
129.5
20
Ø609
SECTION A-A
(SCALE 1:2)
B
20
Ø17.5
DETAIL B
(SCALE 2:1)
Ø36
Ø100
Ø20
Ø60
Ø40
A
A
Ø60
Ø40
Ø36
Ø20
15
20
55
13
3 x 45°
Ø80
185
140
100
20
3 x 45°
30
Ø60
Ø100
SECTION A-A

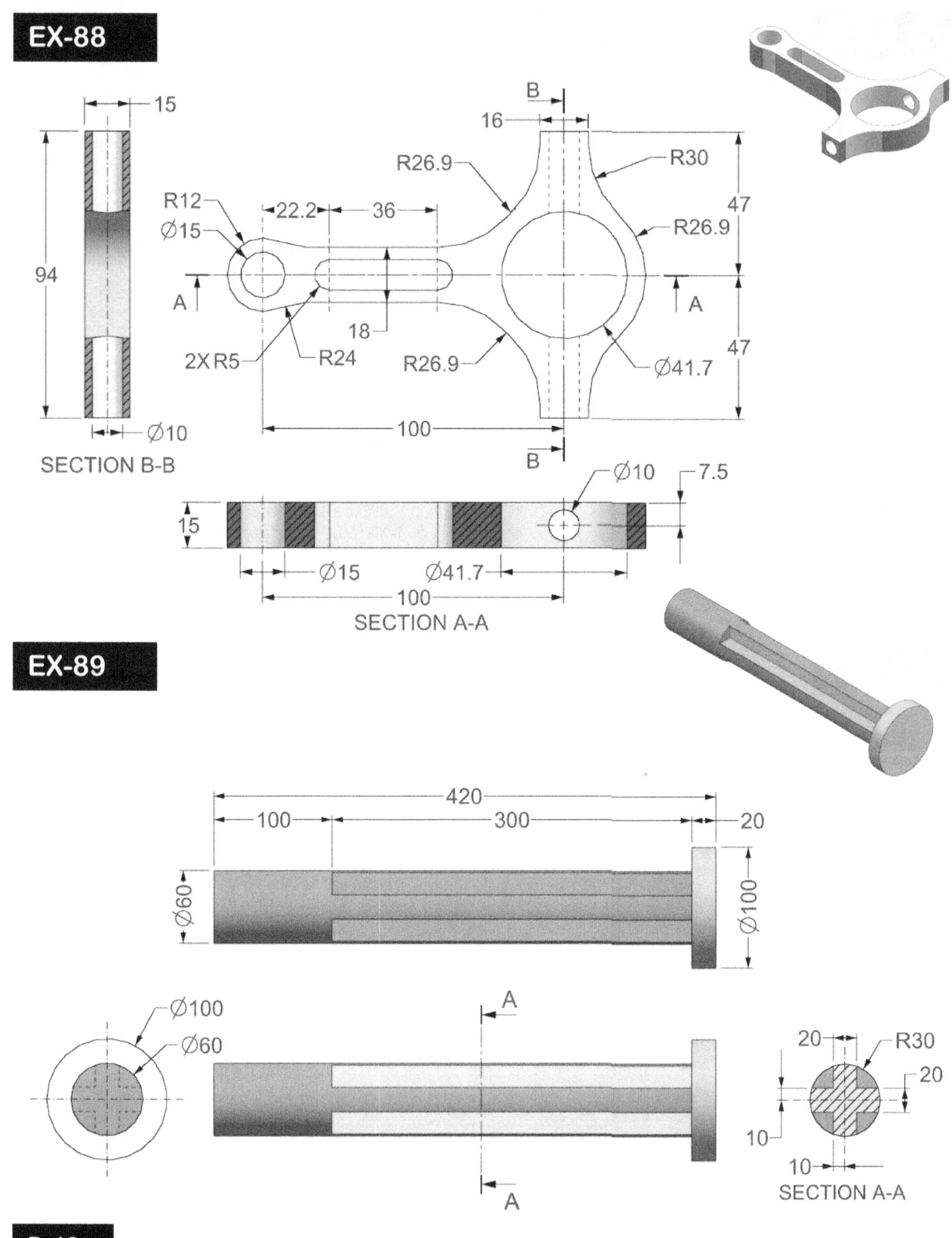

EX-88
15
94
Ø10
SECTION B-B
R12
Ø15
22.2
36
R26.9
16
B
R30
47
R26.9
2X R5
R24
18
R26.9
Ø41.7
100
B
A
A
Ø10
7.5
15
Ø15
Ø41.7
100
SECTION A-A
EX-89
420
100
300
20
Ø60
Ø100
Ø100
Ø60
A
A
20
R30
20
10
10
SECTION A-A
P-46

EX-90
2X Ø32 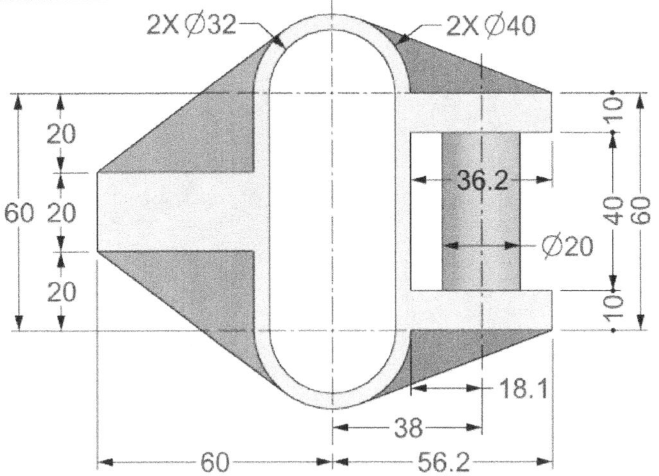
2X Ø40
20
20
60
20
20
36.2
Ø20
10
40
60
10
18.1
38
60
56.2 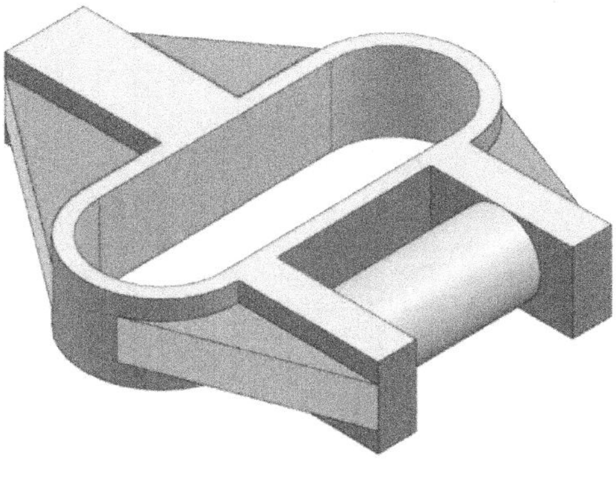
Ø40 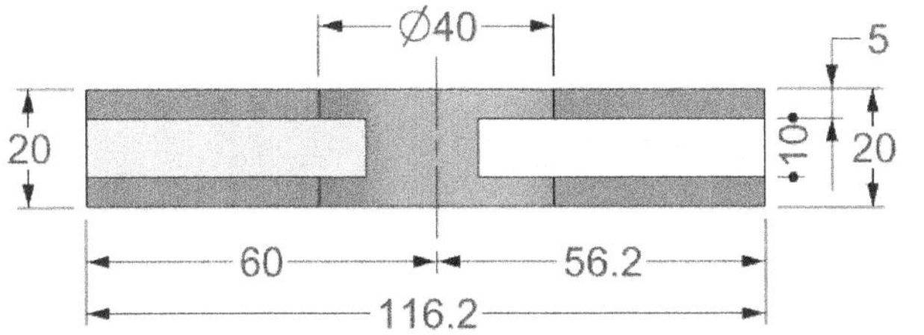
20
20
5
10
60
56.2
116.2
100 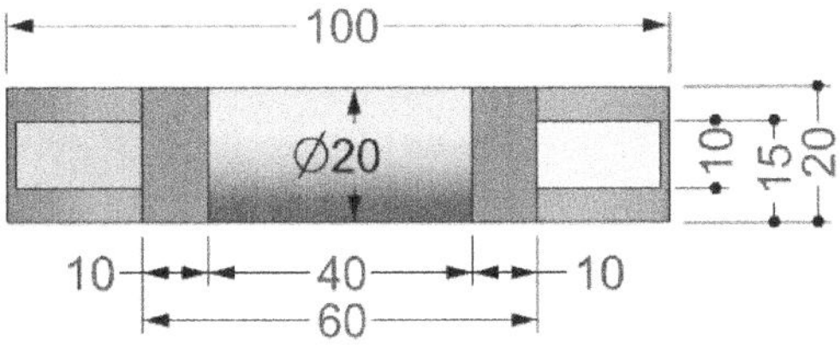
Ø20
20
10
15
10
40
10
10
60
EX-91
30 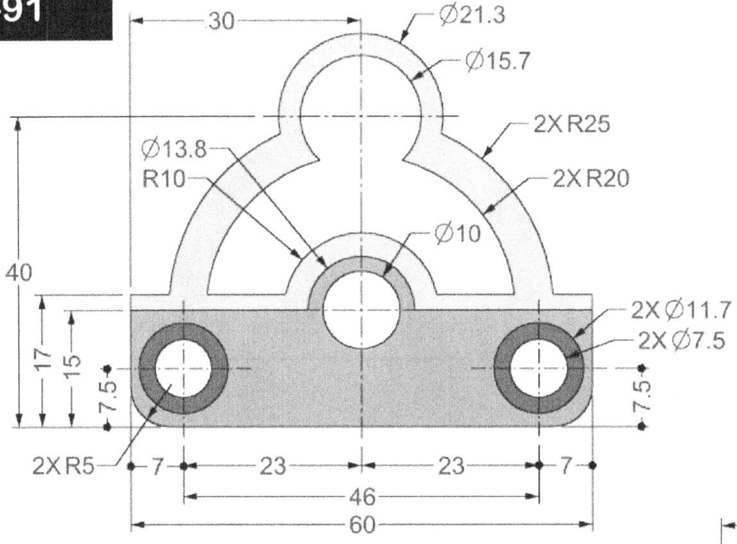
Ø21.3
Ø15.7
2X R25
Ø13.8
R10
2X R20
Ø10
40
2X Ø11.7
2X Ø7.5
17
15
7.5
7.5
2X R5
7
23
23
7
46
60 
Ø11.7 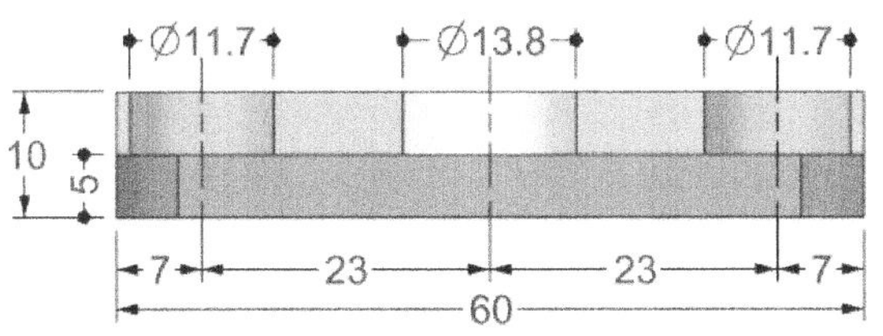
Ø13.8
Ø11.7
10
5
7
23
23
7
60
17 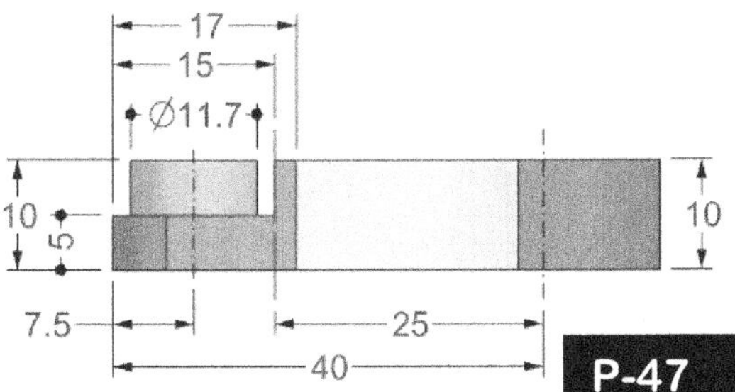
15
Ø11.7
10
5
10
7.5
25
40
P-47

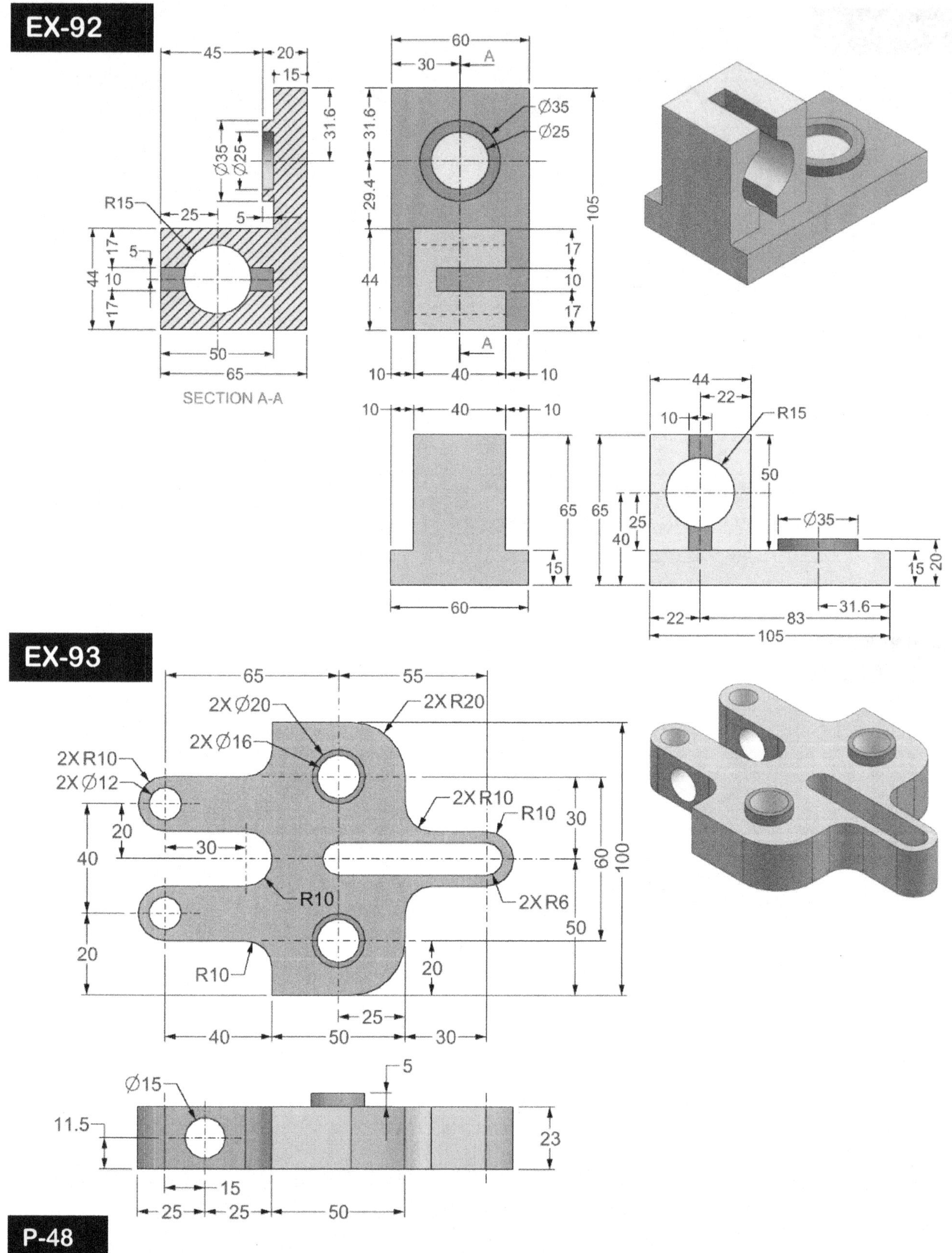

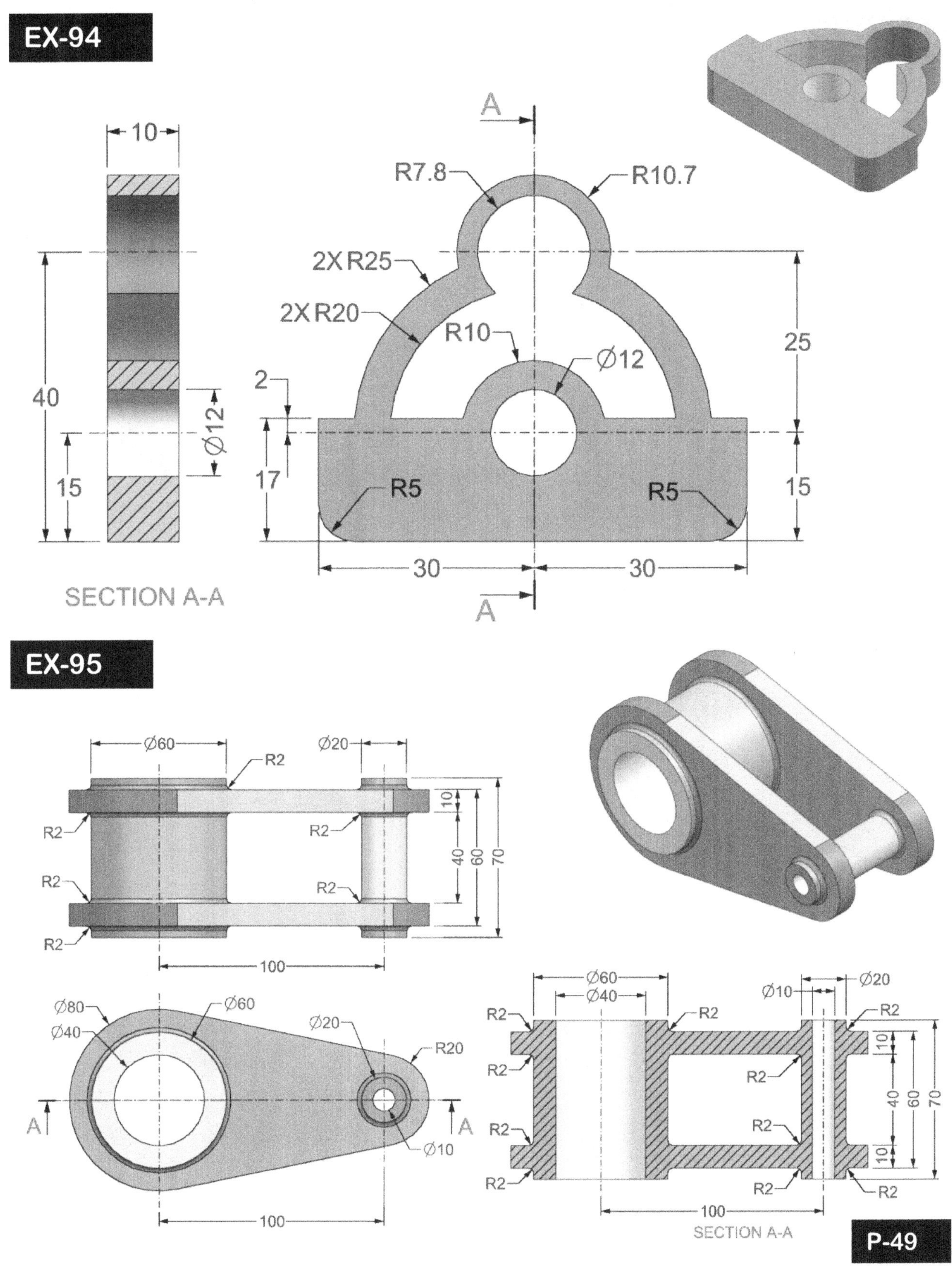

EX-94
SECTION A-A
10
40
15
Ø12
A
A
R7.8
R10.7
2X R25
2X R20
R10
Ø12
2
17
25
15
R5
R5
30
30
EX-95
Ø60
Ø20
R2
R2
R2
R2
R2
R2
10
40
60
70
100
Ø80
Ø60
Ø40
Ø20
R20
A
A
Ø10
100
Ø60
Ø40
Ø10
Ø20
R2
R2
R2
R2
R2
R2
R2
R2
R2
10
40
60
70
10
100
SECTION A-A
P-49

EX-96

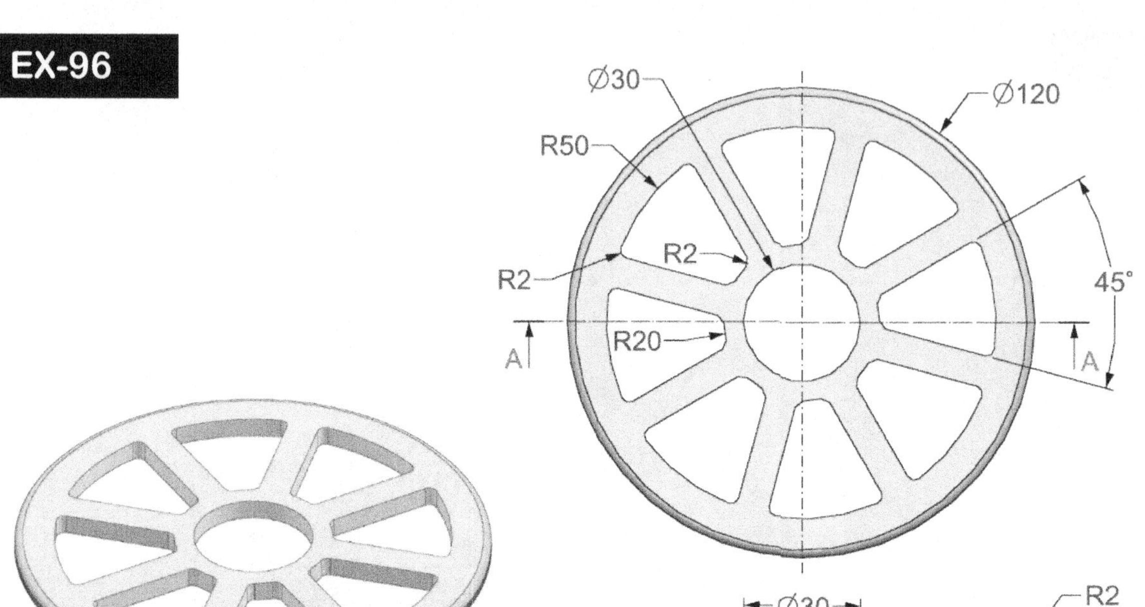

Ø30
Ø120
R50
R2
R2
R20
45°
A
A
Ø30
Ø120
R2
R2
SECTION A-A

EX-97

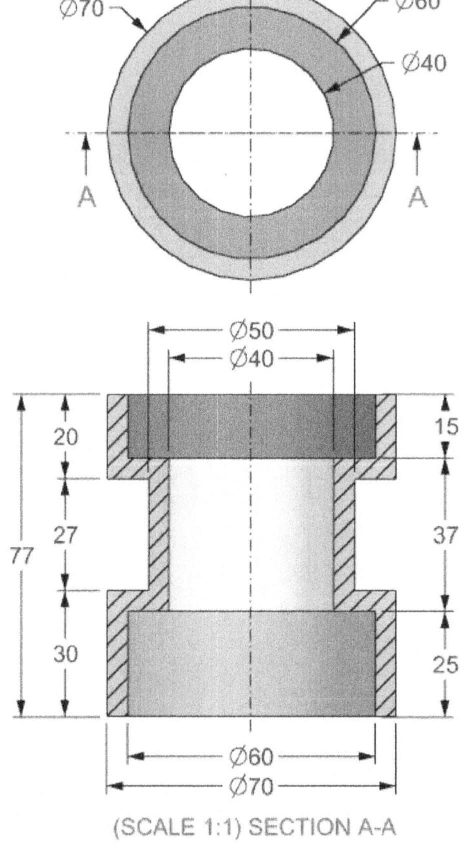

Ø70
Ø60
Ø40
Ø50
Ø40
A
A
20
15
27
37
77
30
25
Ø60
Ø70
(SCALE 1:1) SECTION A-A

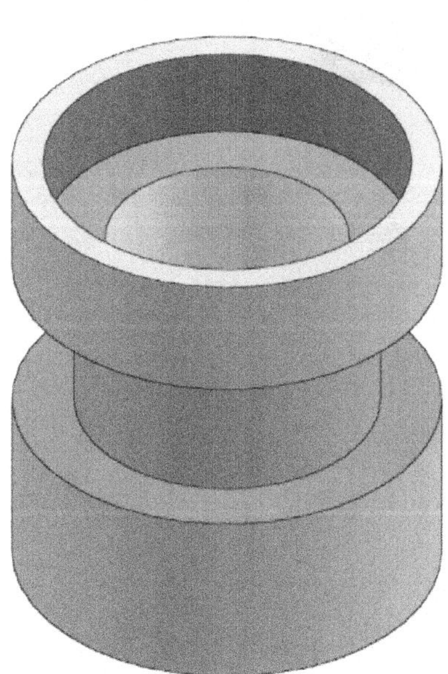

P-50

EX-98
100
20
60
20
5 X 45°
15
15
4X Ø16
30
170
140
50
15
15
70
30
15
15
100
15
70
15
70
60
100
60
50
60
30
30
30
30
70
60
60
10
36.2
30
37.6
30
36.2
10
EX-99
ON PCD Ø41
Ø2
A
Ø36
Ø46
Ø16
Ø46
Ø36
Ø16
Ø36
Ø36
5
5
5
20
5
5
40
SECTION A-A
A
Ø46
Ø36
Ø2
5
5
20
40
5
5
Ø36
Ø46
P-51

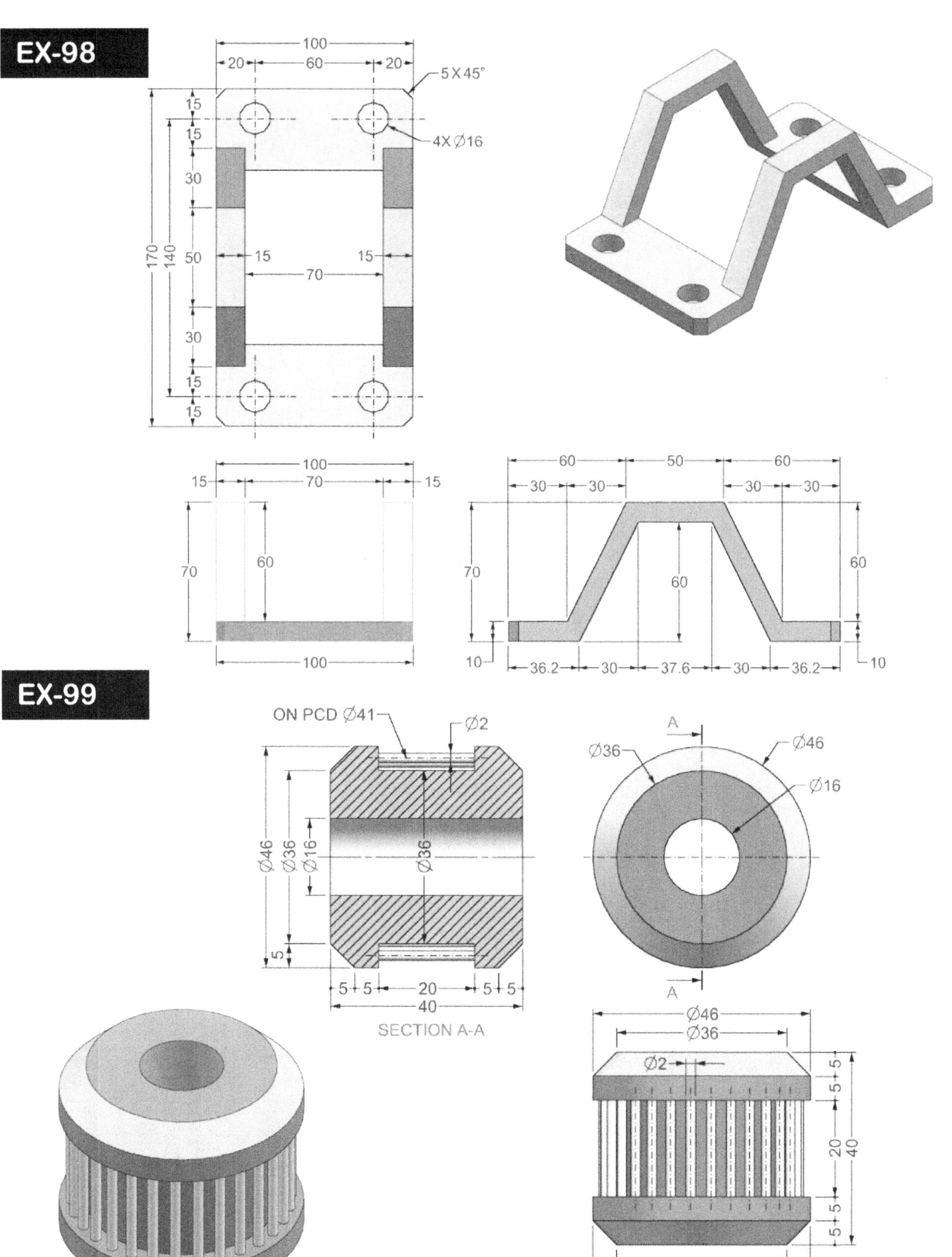

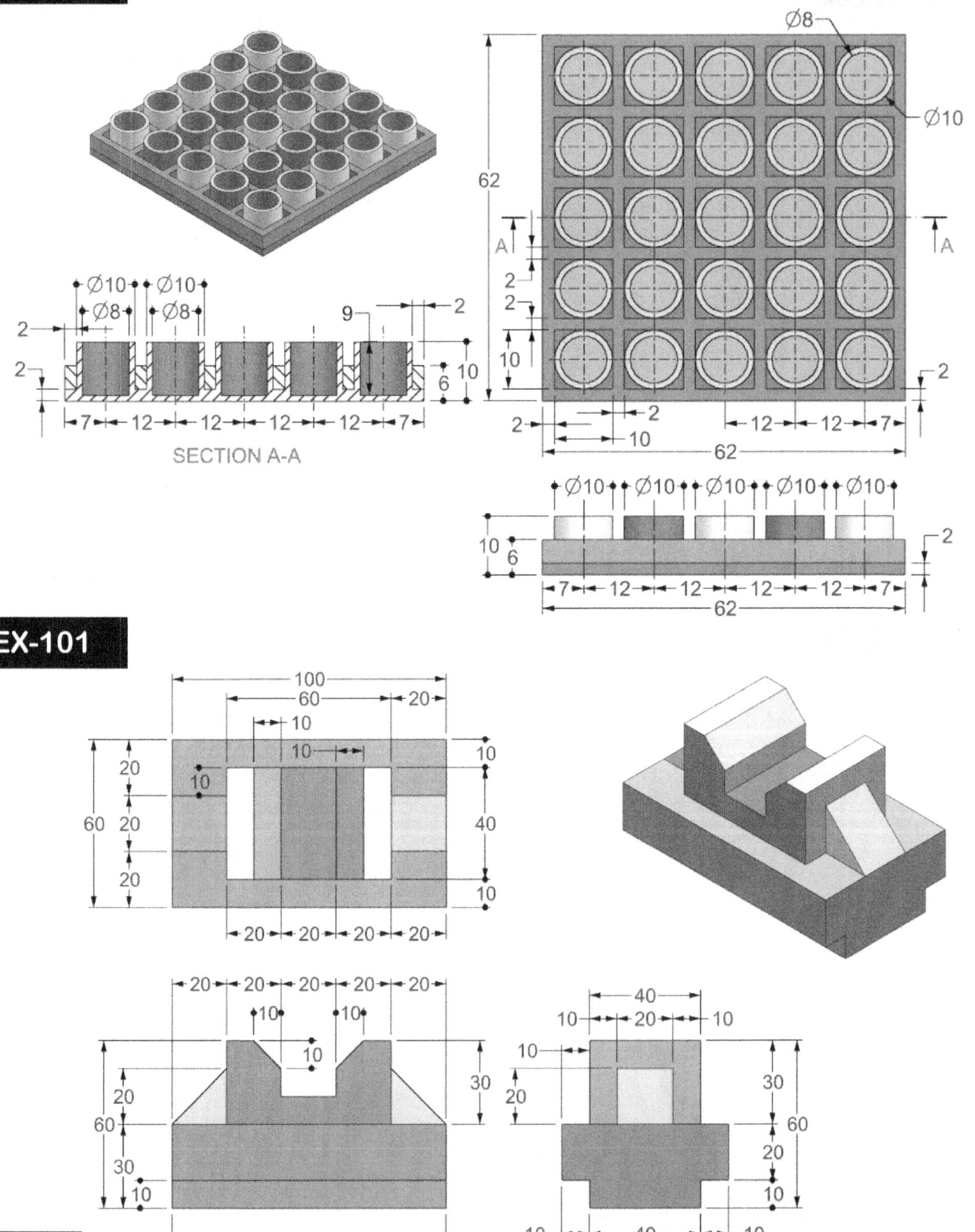

EX-100
∅8
∅10
62
A
A
2
2
10
2
2
10
62
12 12 7
∅10 ∅10
∅8 ∅8
2
2 9
2
10
6
7 12 12 12 12 7
SECTION A-A
∅10 ∅10 ∅10 ∅10 ∅10
10
6
2
7 12 12 12 12 7
62
EX-101
100
60
20
10
10
20
10
10
60
20
40
20
10
20 20 20 20
20 20 20 20 20
10 10
10
20
30
60
20
30
10
100
40
10 20 10
10
10
20
30
30
60
20
10
10 40 10
60
P-52

EX-102

124
14
96
14
8
4X Ø20
4X Ø16
124
96
14
14
8
8
8
Ø20
Ø20
7.5
8
23
8
14
96
14
124
Ø20
Ø20
8
7.5
14
96
14
124

EX-103

3
7
7
10
10
10
10
23
23
23
23
107
5
3
23
23
23
23
113
10
10
3
7
16
8
113
10
10
5
3
16
8
107

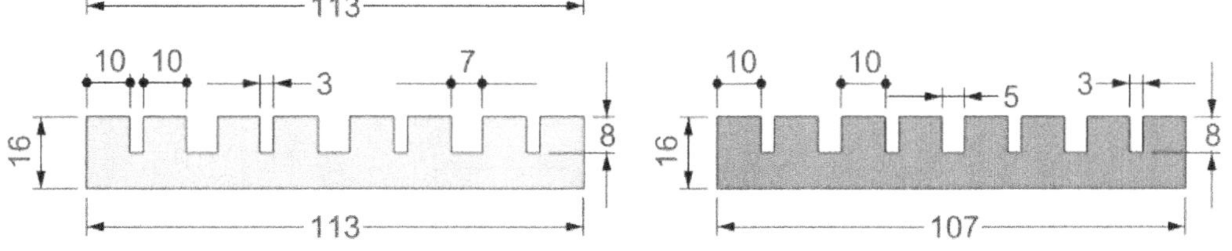

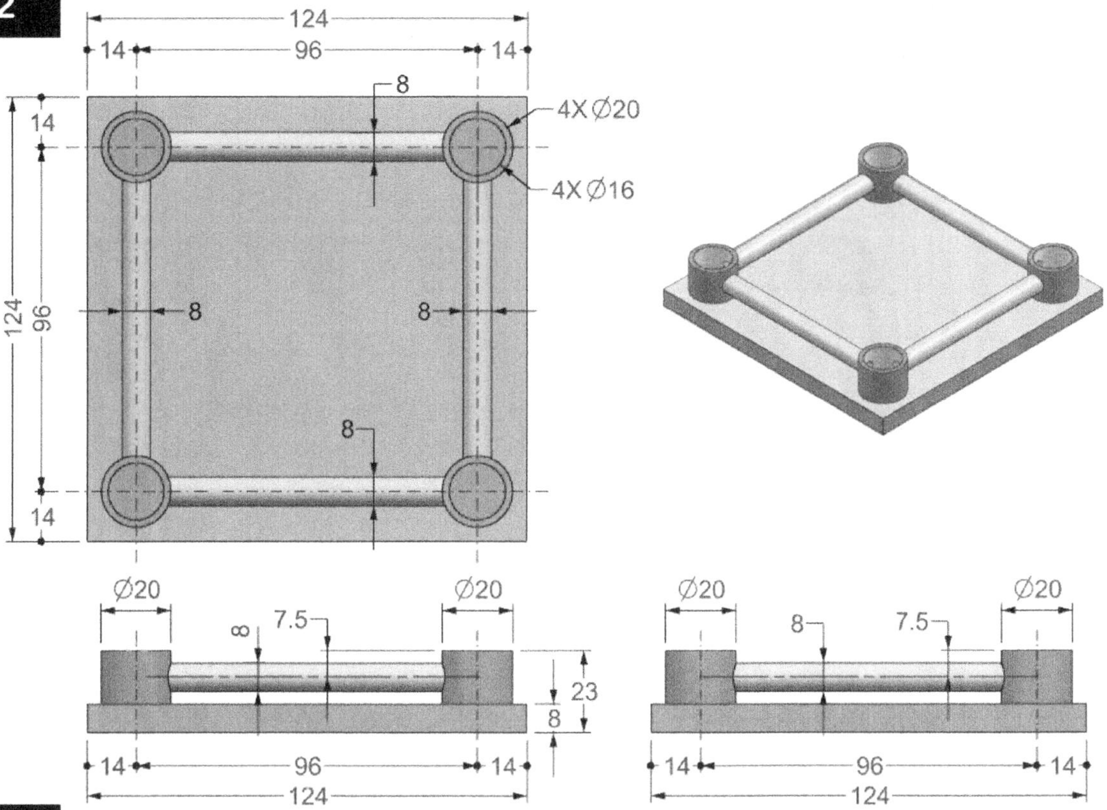

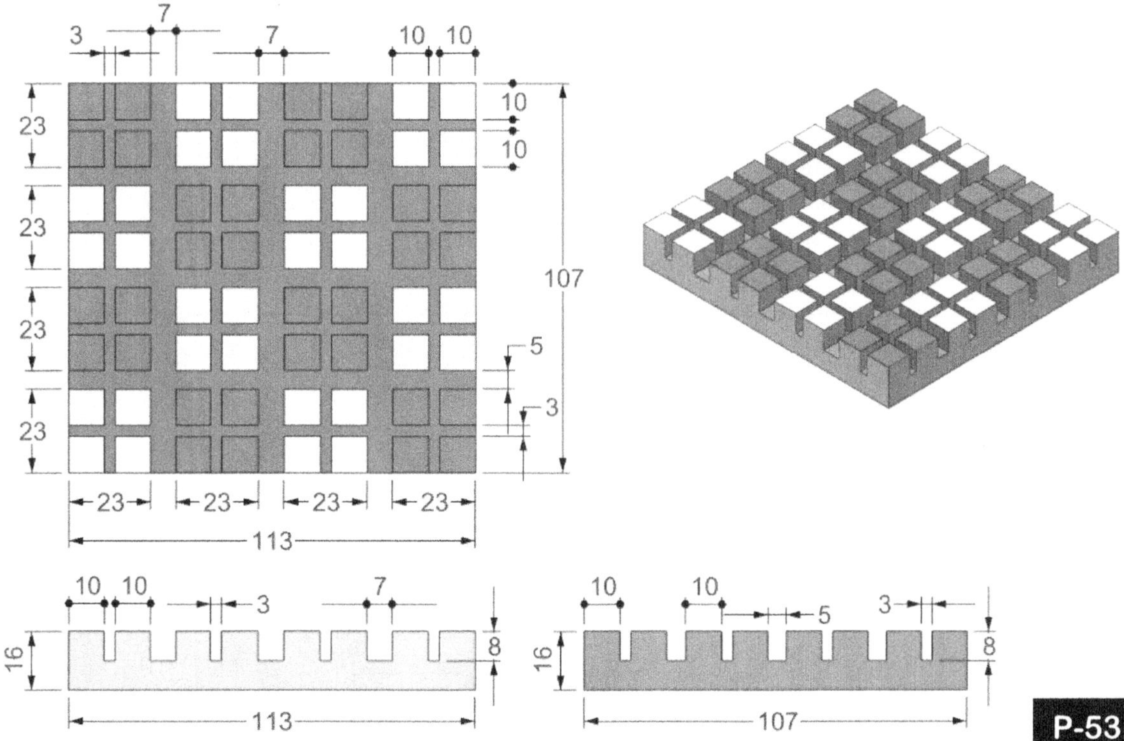

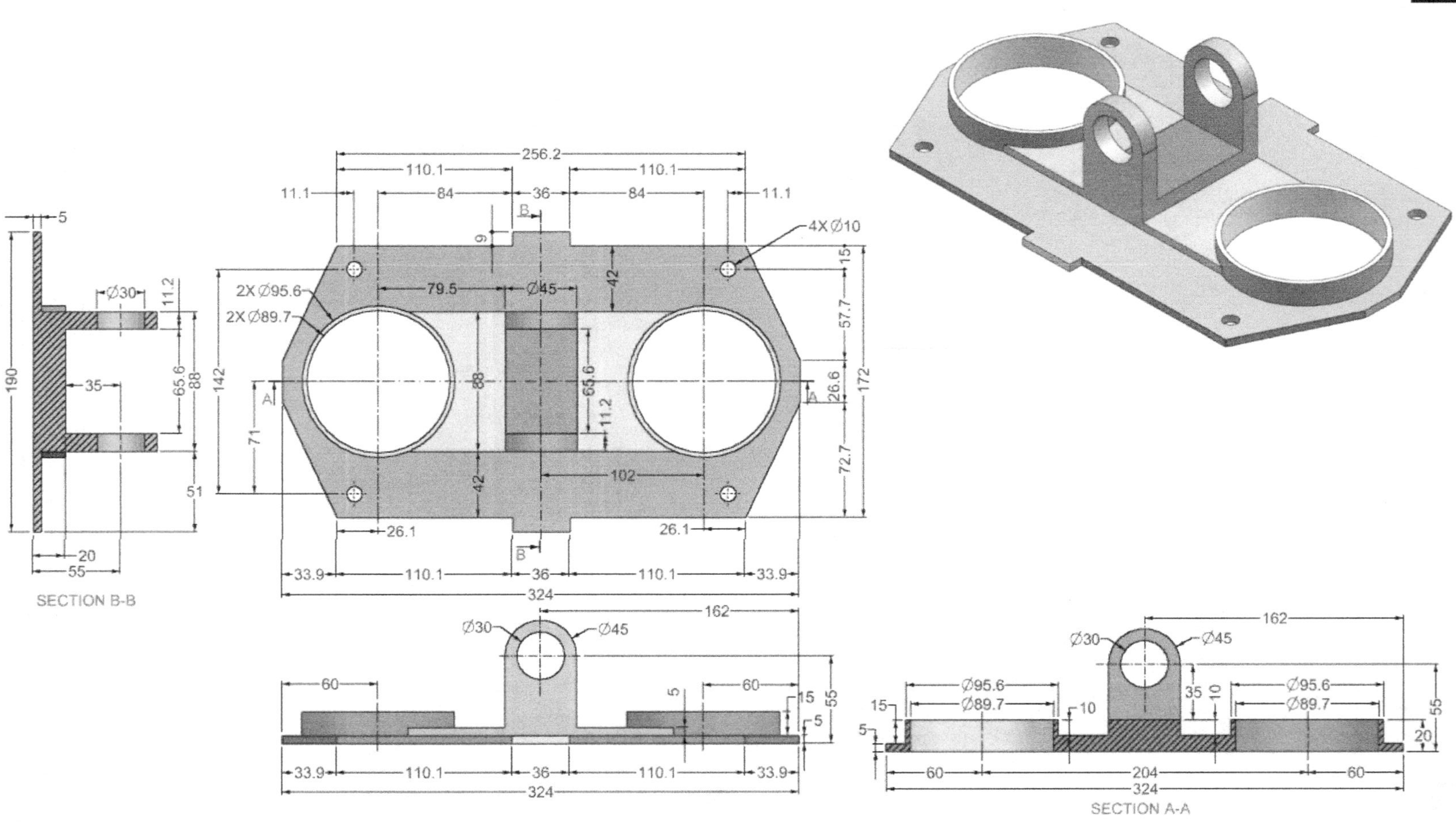

SECTION B-B
SECTION A-A
4X ⌀10
2X ⌀95.6
2X ⌀89.7
⌀45
⌀30
⌀95.6
⌀89.7

EX-105
324 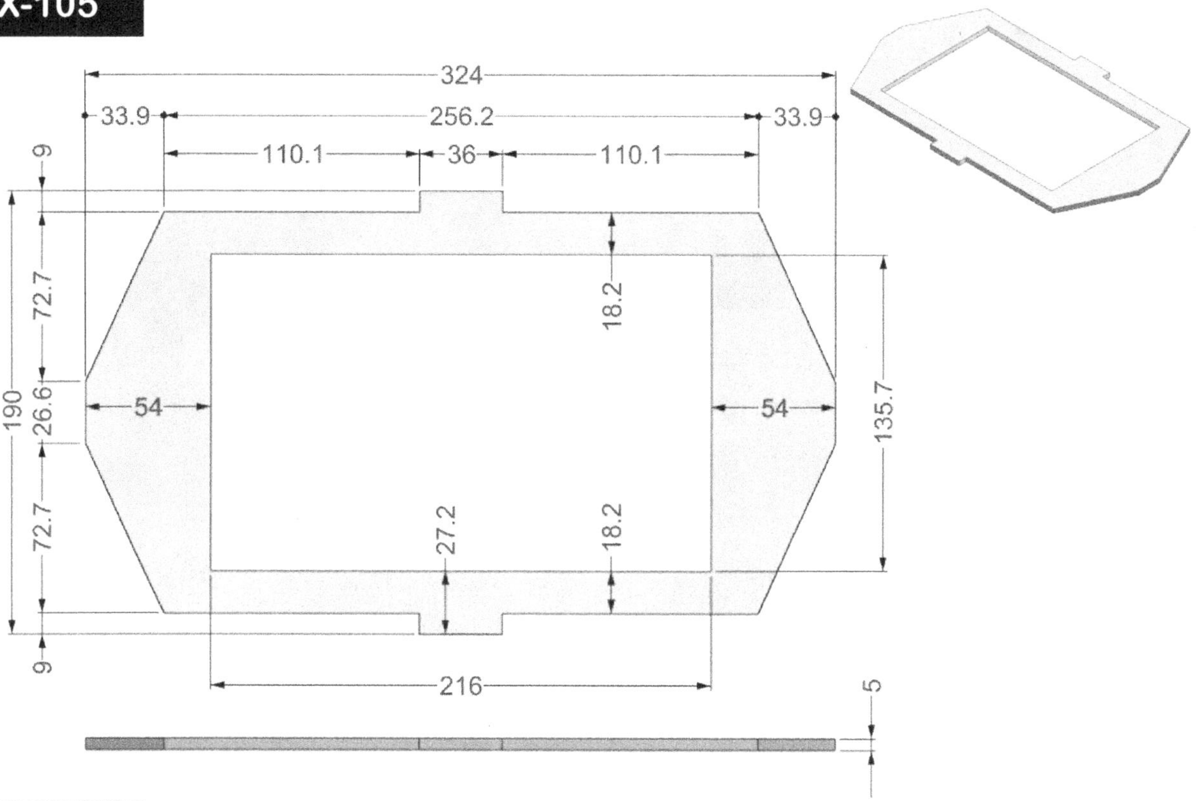
33.9
256.2
33.9
9
110.1
36
110.1
72.7
18.2
190
26.6
54
54
135.7
72.7
27.2
18.2
9
216
5

EX-106
7
Ø8
Ø12
A
2X Ø12
2X Ø8
12.5
2X R4
70
56.8
70
56.8
31.6
4
5
12.7
2X R64.5
6.6
A
10
SECTION A-A
A
20
P-55

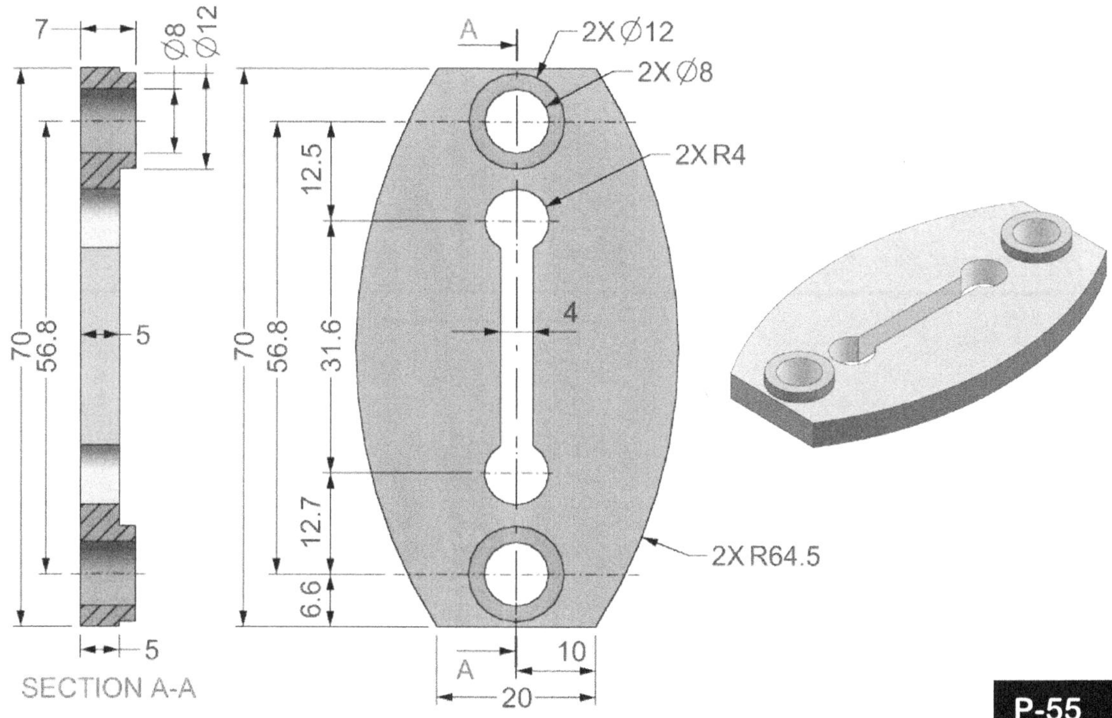

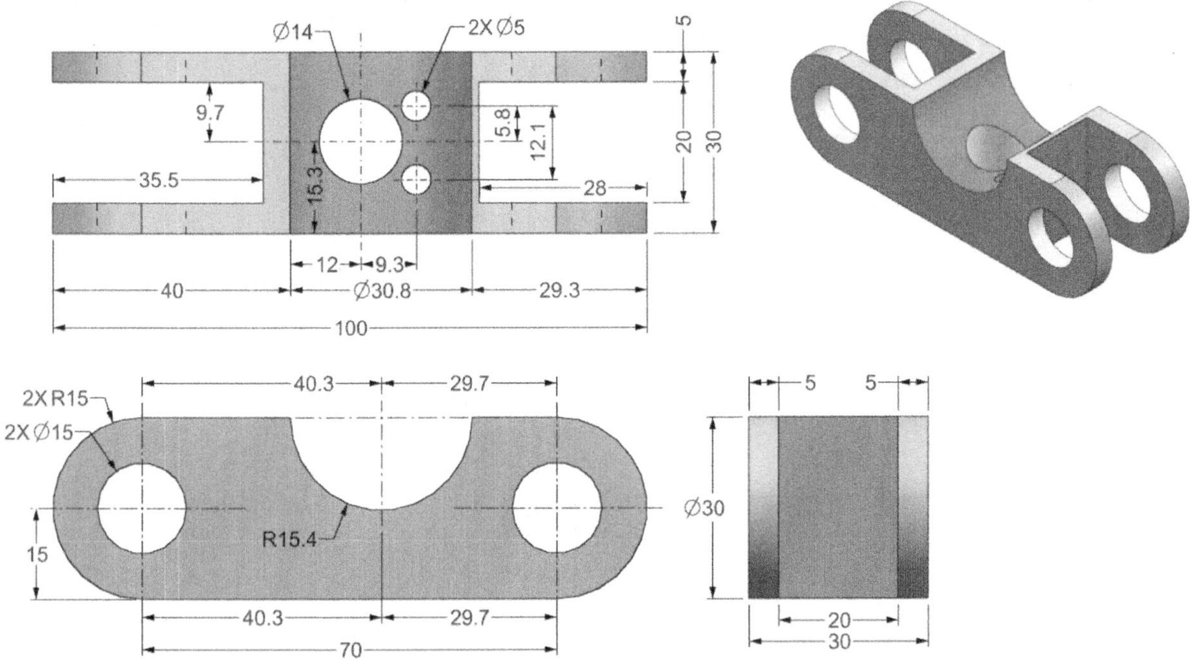

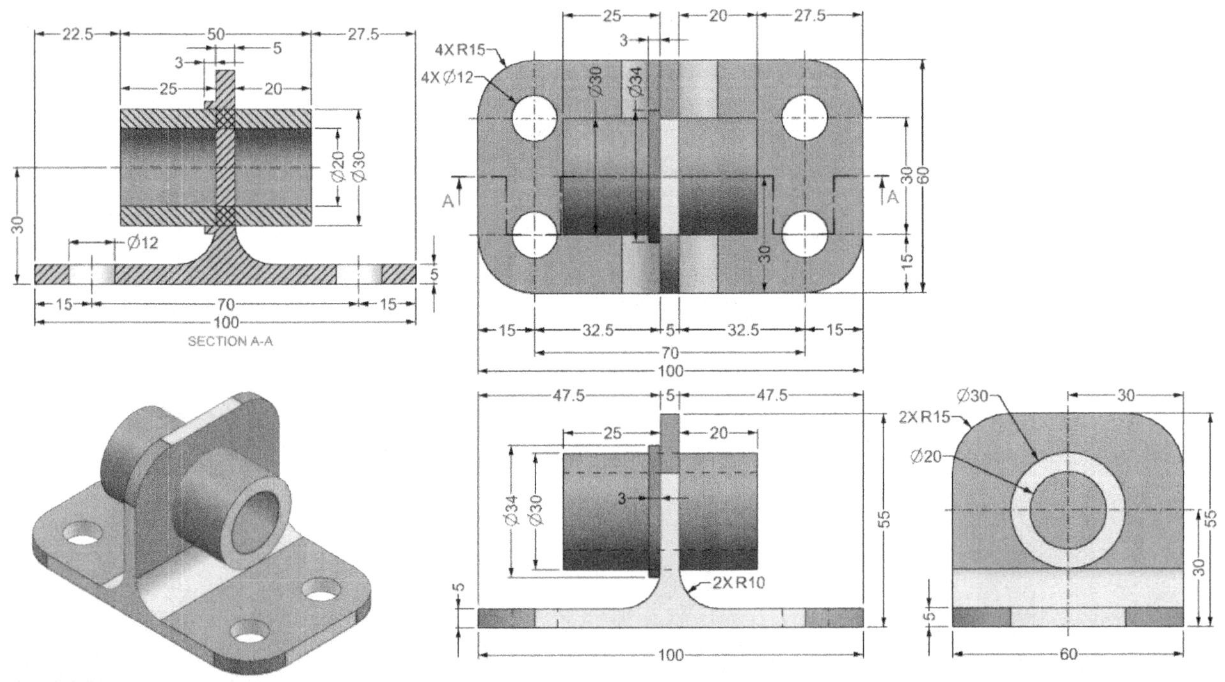

EX-109

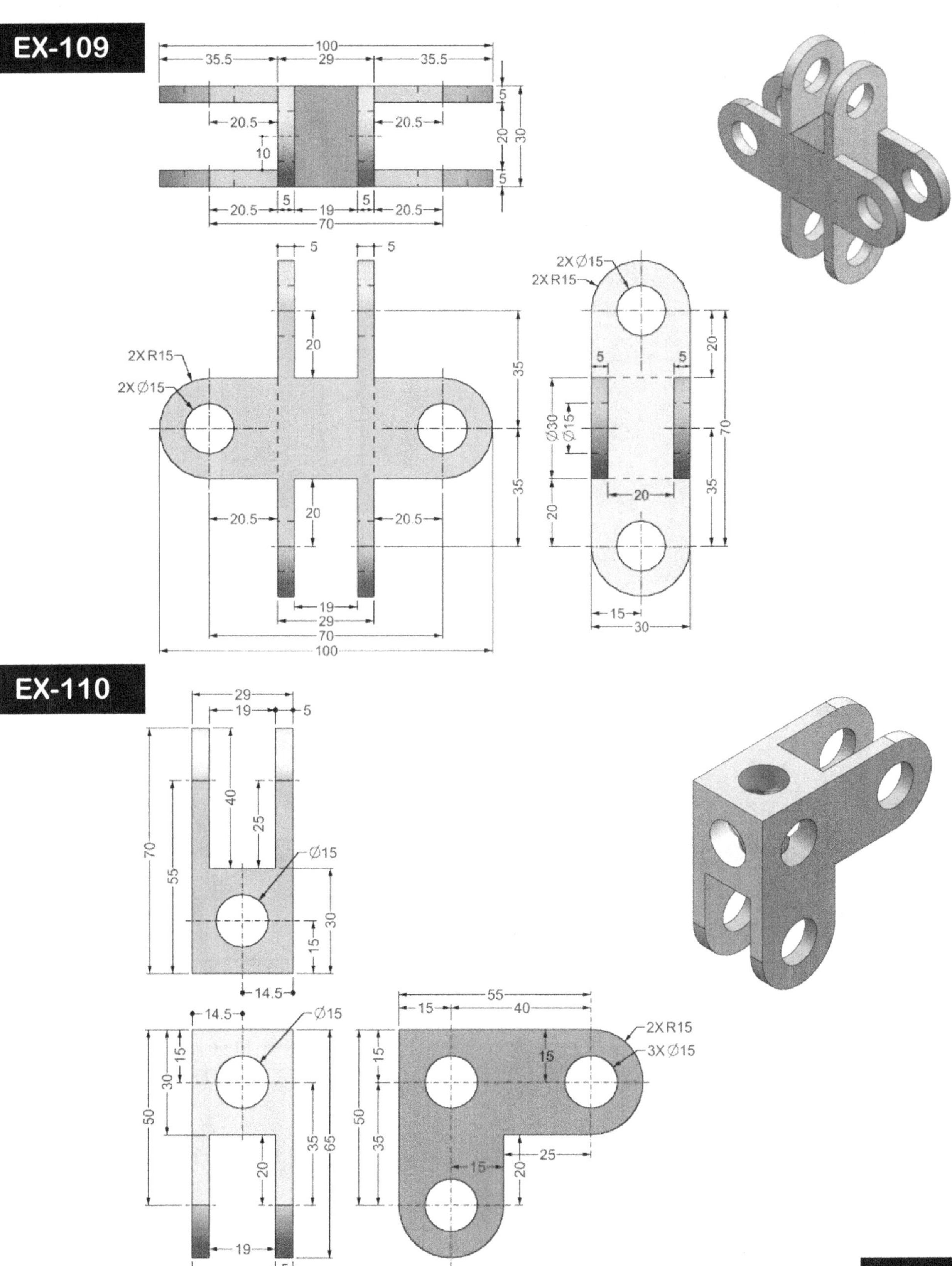
EX-109
100
35.5
29
35.5
5
20
30
20.5
20.5
10
5
5
20.5
5
19
5
20.5
70
5
5
20
2XR15
2XØ15
35
35
20.5
20
20.5
19
29
70
100
2XØ15
2XR15
5
5
20
Ø30
Ø15
70
35
20
20
15
30
EX-110
29
19
5
40
25
70
55
Ø15
30
15
14.5
14.5
Ø15
15
30
50
35
65
20
19
5
29
55
15
40
15
15
50
35
2XR15
3XØ15
25
15
20
P-57

EX-111
Ø28 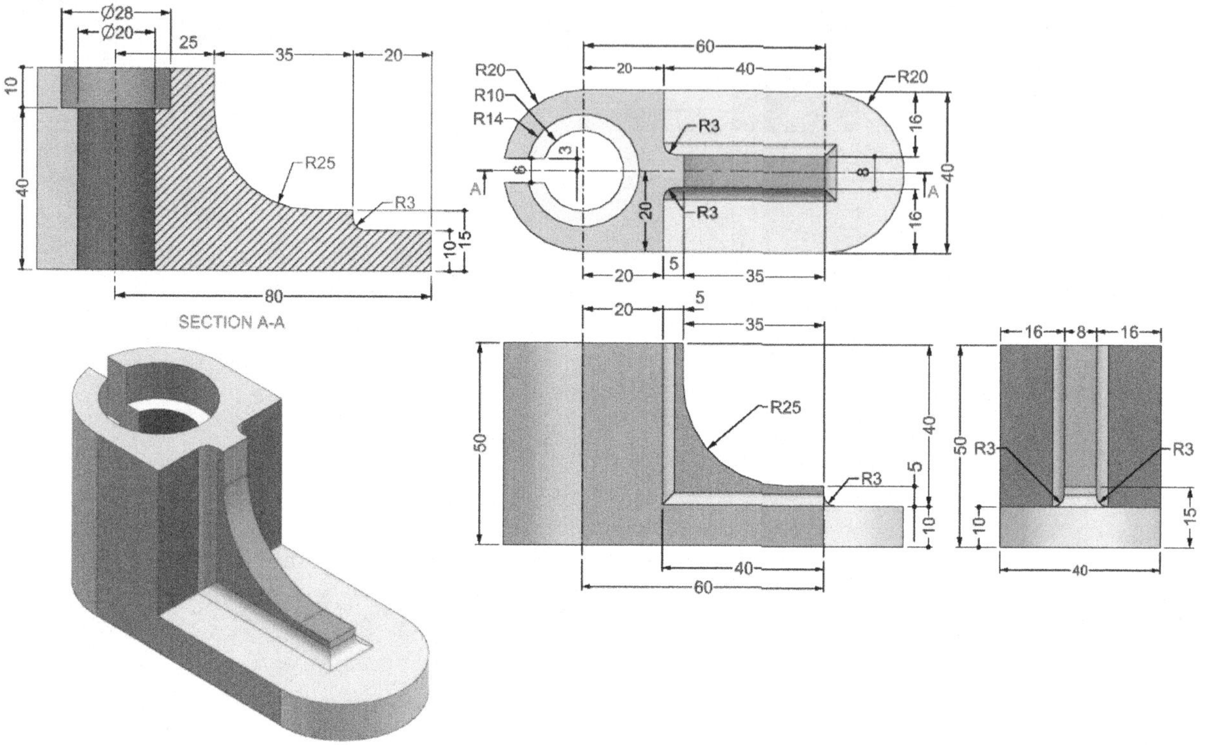
Ø20
25
35
20
10
40
R25
R3
10
15
80
SECTION A-A
60
20
40
R20
R10
R14
R20
16
R3
3
6
8
40
20
A
A
R3
16
20
5
35
20
5
35
50
R25
R3
40
5
10
40
60
16
8
16
50
R3
R3
10
15
40
EX-112
2XR10
2XR15
4XØ14
4XØ12
15
10
10
20
10
10
15
10
17.5
32.5
32.5
17.5
65
100
35
Ø30
35
40
Ø14
10
12
17.5
32.5
32.5
17.5
65
100
20
10
20
30
Ø14
Ø14
10
12
10
30
10
50
P-58

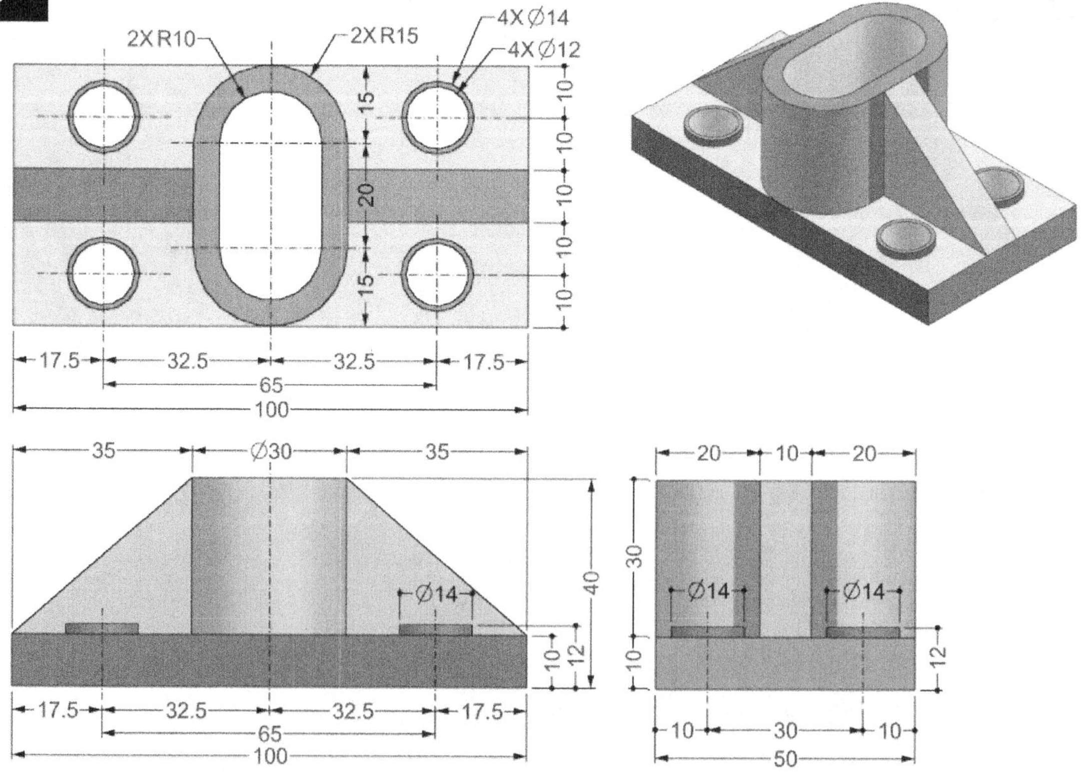

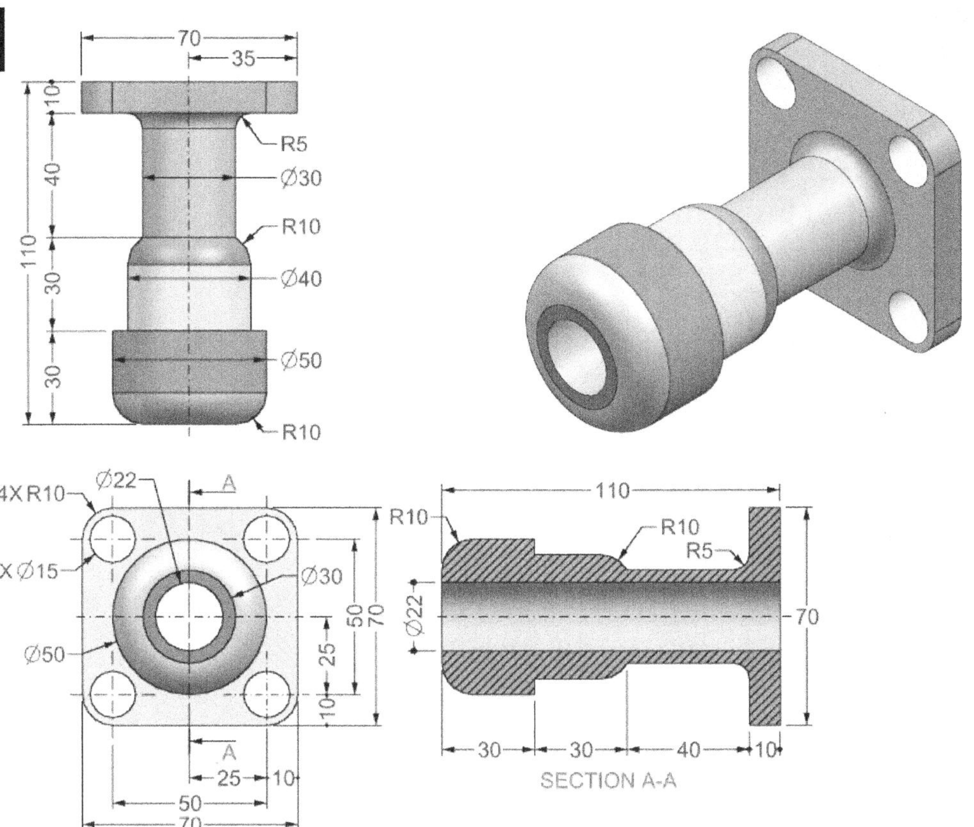

70
35
10
40
110
30
30
R5
Ø30
R10
Ø40
Ø50
R10
4X R10
Ø22
A
4X Ø15
Ø30
Ø50
50
70
25
10
A
25
10
50
70
110
R10
R10
R5
Ø22
70
30
30
40
10
SECTION A-A

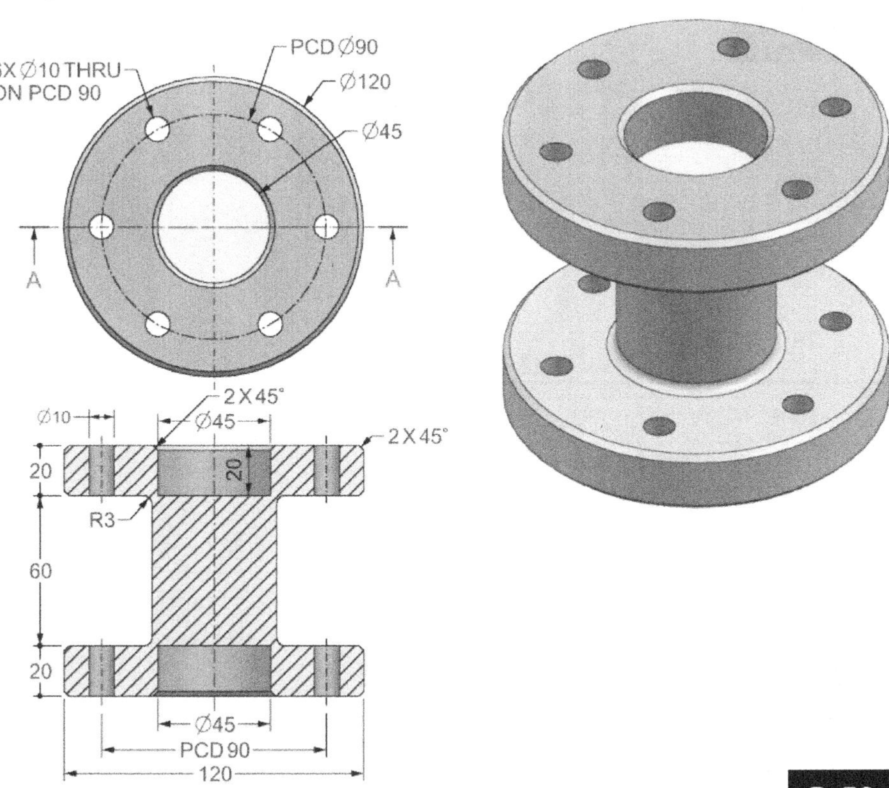

6X Ø10 THRU
ON PCD 90
PCD Ø90
Ø120
Ø45
A
A
2 X 45°
Ø10
Ø45
2 X 45°
20
20
R3
60
20
Ø45
PCD 90
120
SECTION A-A

EX-115

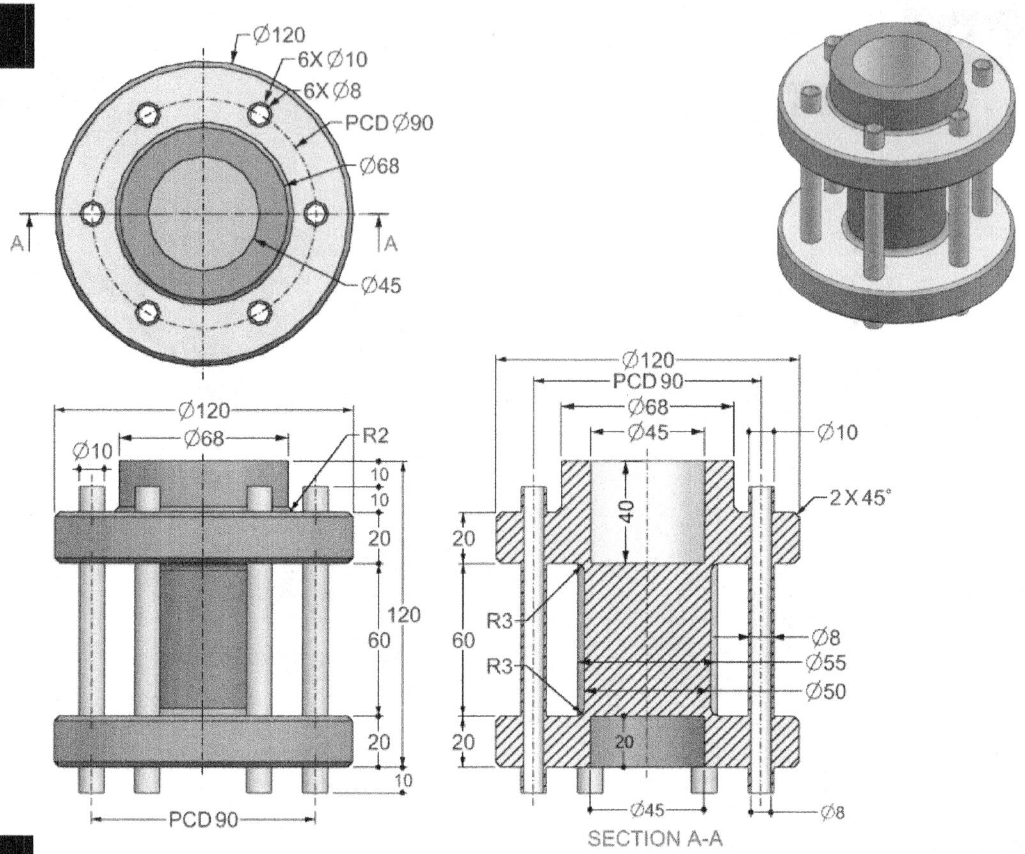

Ø120
6X Ø10
6X Ø8
PCD Ø90
Ø68
Ø45
A
A
Ø120
Ø68
Ø10
R2
10
10
20
120
60
20
10
PCD 90
Ø120
PCD 90
Ø68
Ø45
Ø10
2 X 45°
40
20
60
20
R3
R3
Ø8
Ø55
Ø50
20
Ø45
Ø8
SECTION A-A

EX-116

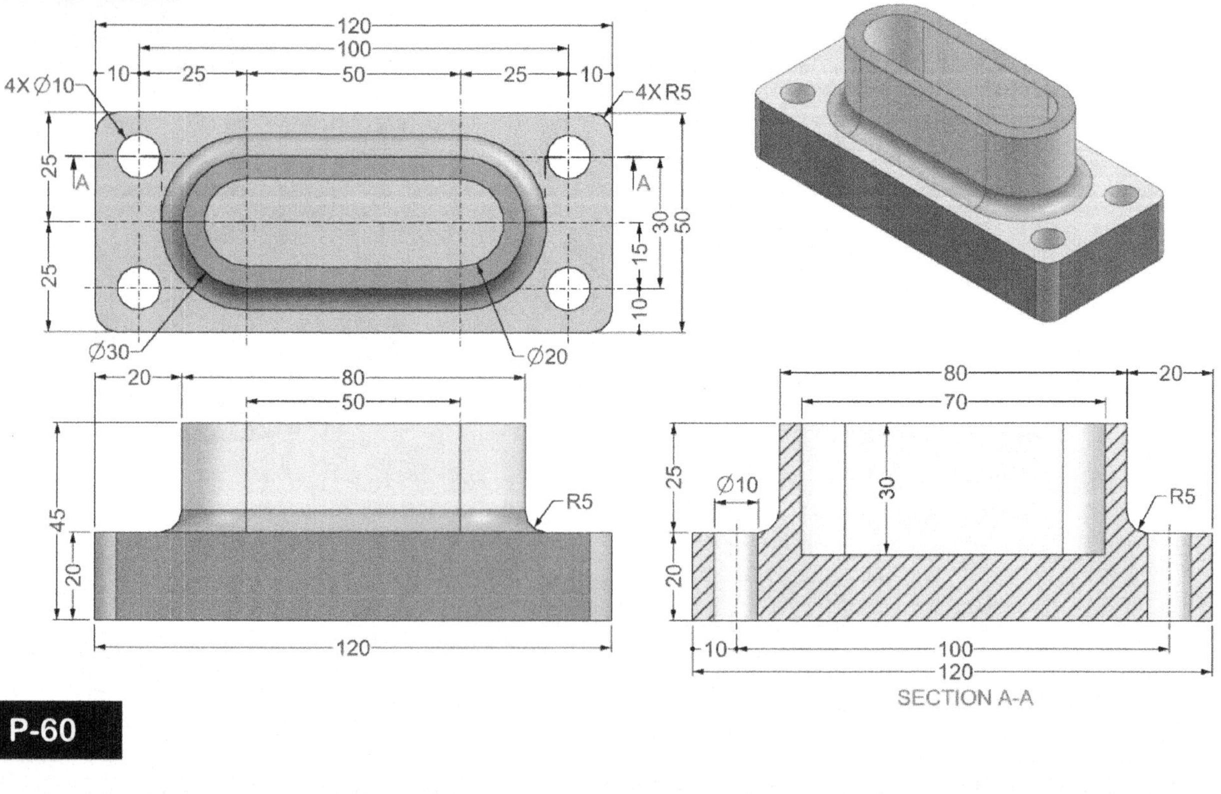

120
100
10
25
50
25
10
4X Ø10
4X R5
25
25
A
A
30
50
15
10
Ø30
Ø20
20
80
50
45
20
R5
120
80
70
20
25
Ø10
30
R5
20
10
100
120
SECTION A-A

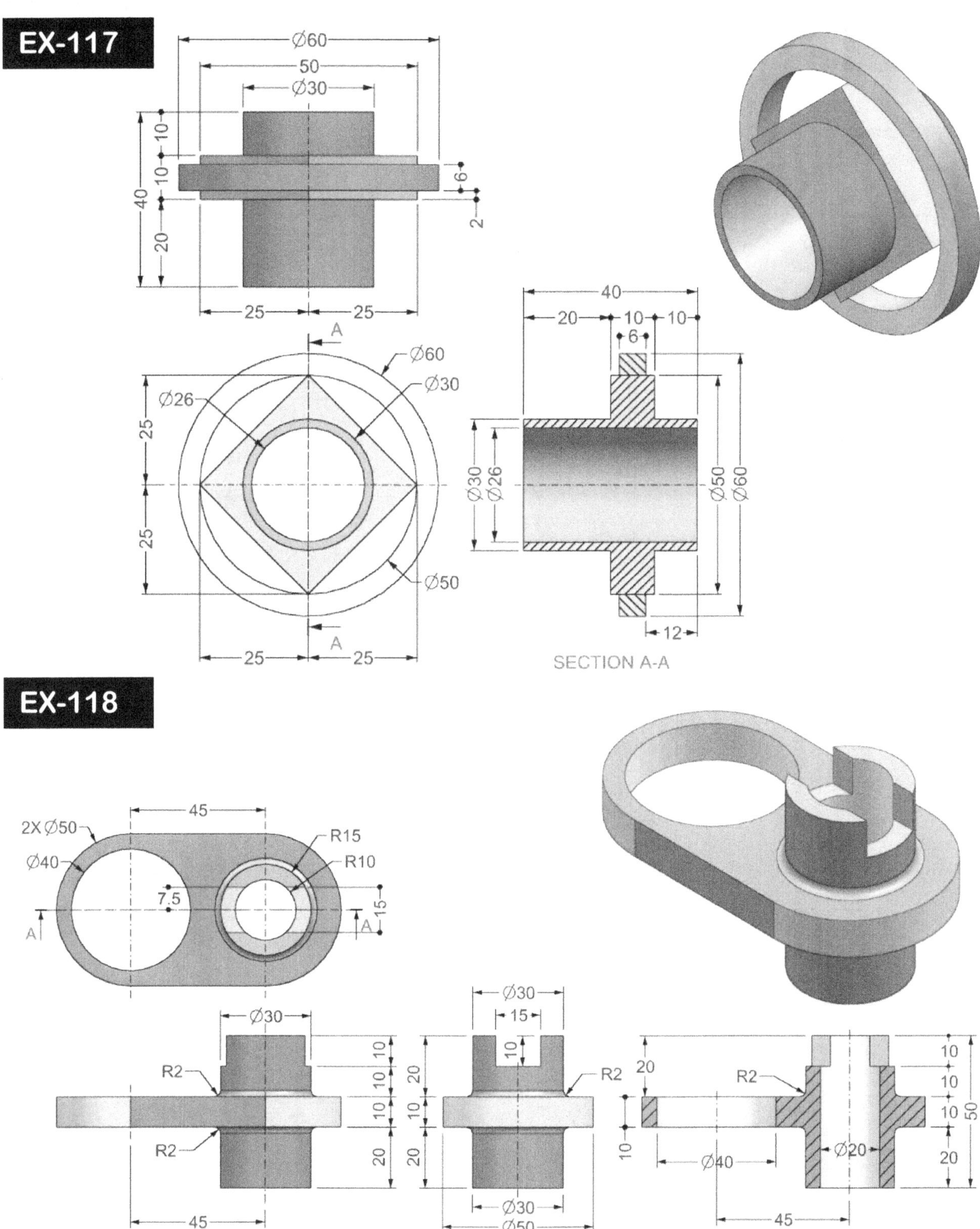

EX-117
Ø60
50
Ø30
10
10
40
6
20
2
25
25
A
A
Ø60
Ø30
Ø26
25
25
25
25
Ø50
25
25
40
20
10
10
6
Ø30
Ø26
Ø50
Ø60
12
SECTION A-A
EX-118
45
2X Ø50
R15
Ø40
R10
7.5
15
A
A
Ø30
Ø30
Ø30
15
R2
10
10
10
10
10
10
20
10
20
R2
20
20
R2
20
R2
Ø40
Ø20
10
10
10
50
20
45
Ø30
Ø50
45
SECTION A-A
P-61

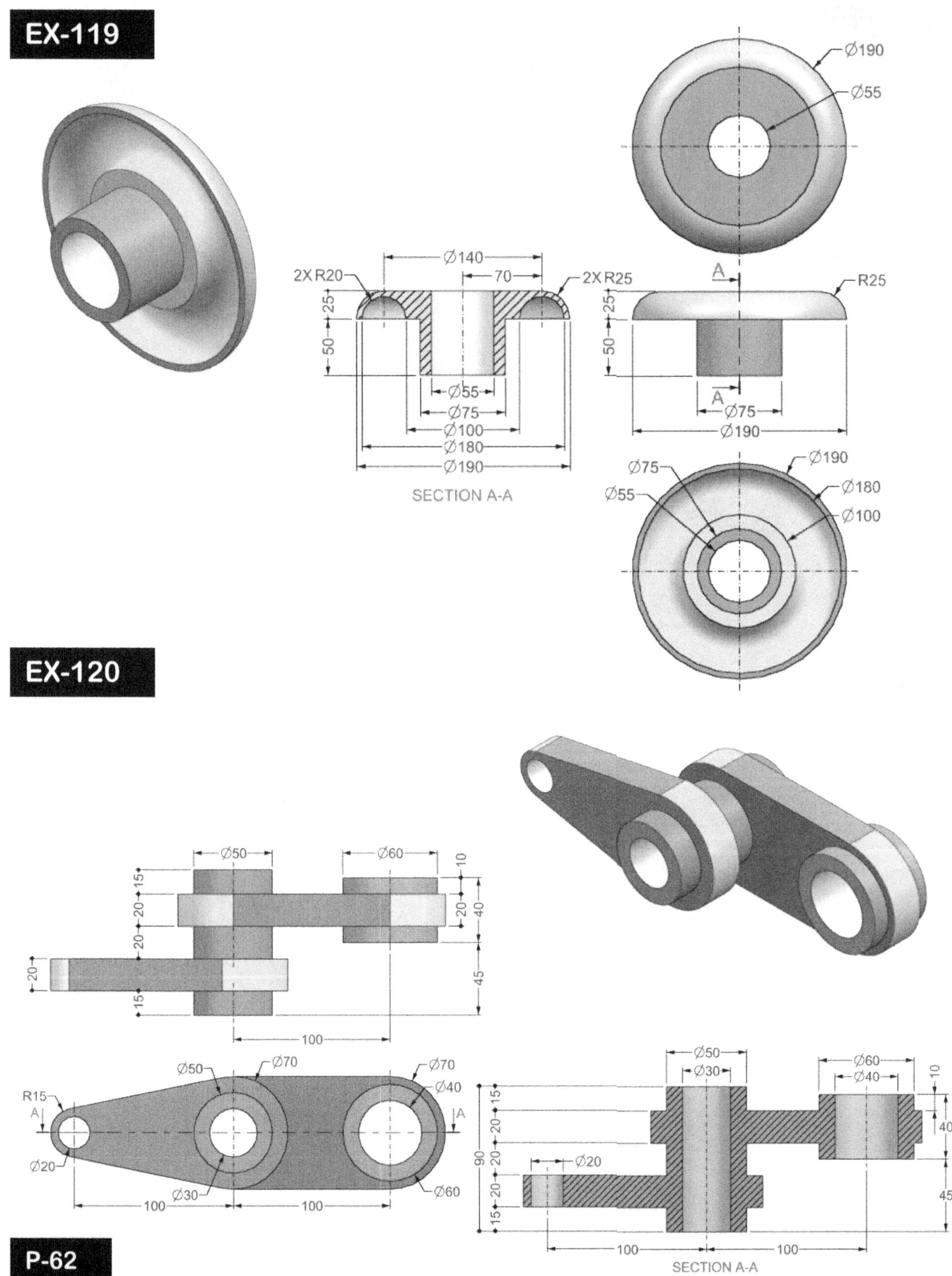
EX-119
Ø190
Ø55
Ø140
70
2X R20
2X R25
25
50
Ø55
Ø75
Ø100
Ø180
Ø190
SECTION A-A
A
R25
25
50
A
Ø75
Ø190
Ø75
Ø190
Ø55
Ø180
Ø100
EX-120
Ø50
Ø60
10
15
20
20
20
20
40
20
15
20
100
R15
Ø50
Ø70
Ø70
Ø40
A
A
Ø20
Ø30
Ø60
100
100
Ø50
Ø30
Ø60
Ø40
10
20
15
20
20
90
20
Ø20
40
15
100
100
45
SECTION A-A
P-62

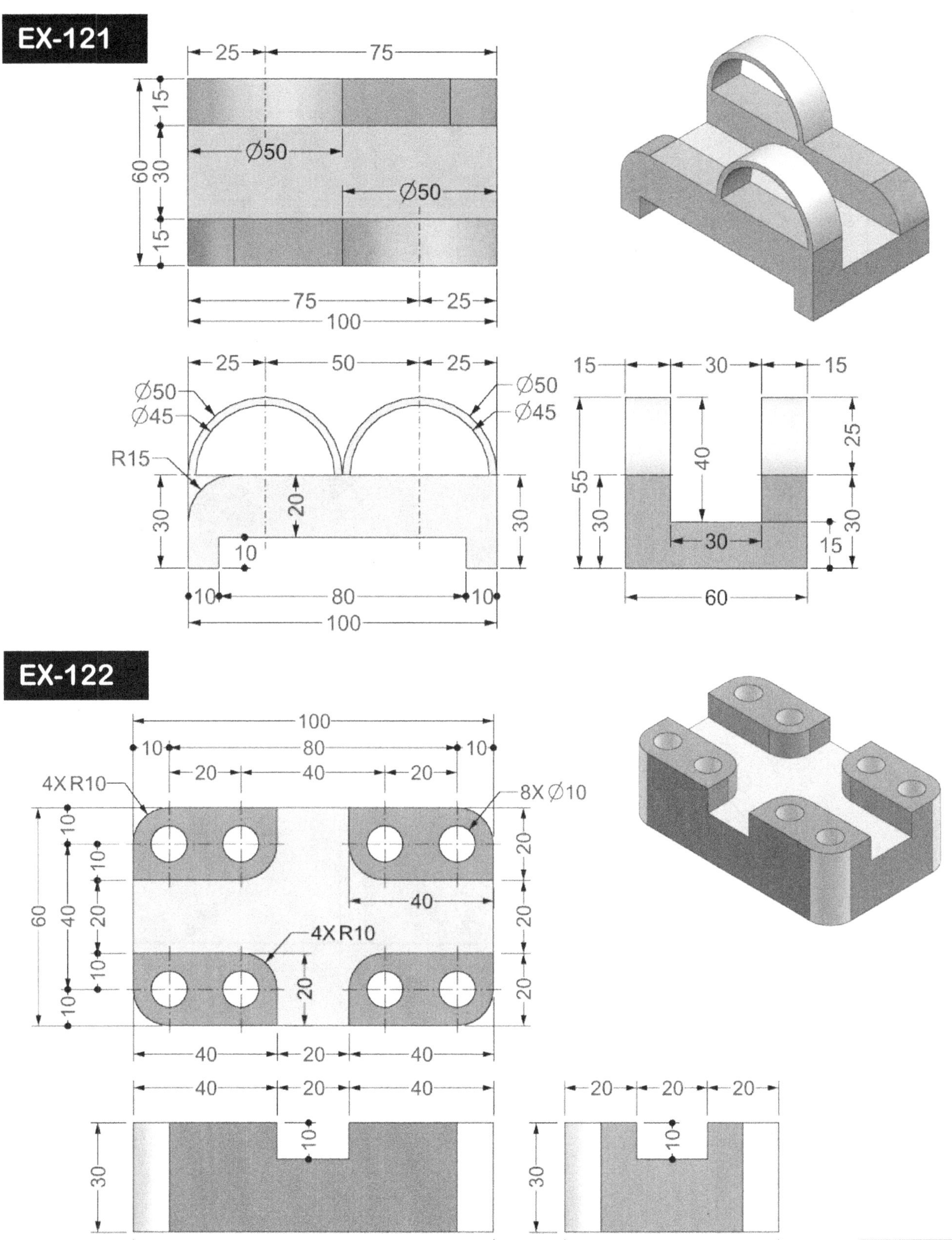

EX-121
25
75
15
60
30
15
Ø50
Ø50
75
25
100
25
50
25
Ø50
Ø45
Ø50
Ø45
R15
30
20
30
10
10
80
10
100
15
30
15
55
40
25
30
30
30
15
60
EX-122
100
10
80
10
20
40
20
4X R10
8X Ø10
60
10
40
10
20
10
10
40
20
20
4X R10
20
40
20
40
40
20
40
40
20
40
30
10
100
20
20
20
30
10
60
P-63

EX-123
10 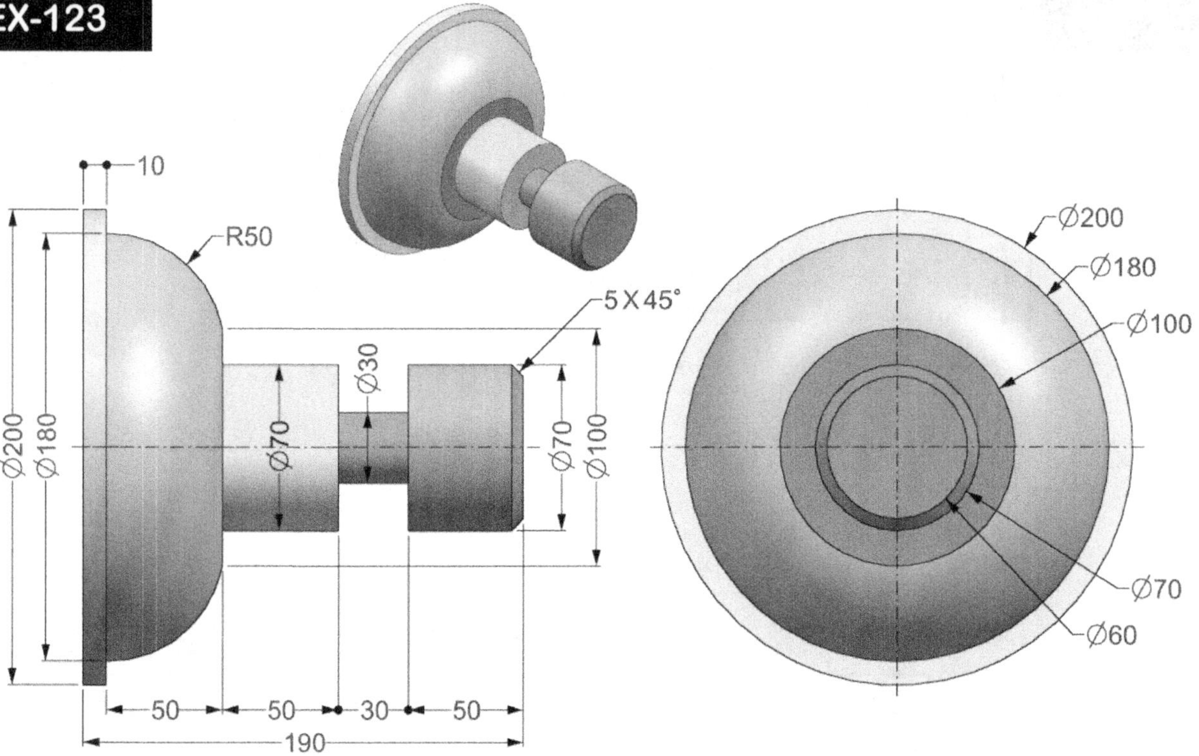
R50
5 X 45°
Ø30
Ø70
Ø70
Ø100
Ø200
Ø180
50
50
30
50
190
Ø200
Ø180
Ø100
Ø70
Ø60

EX-124
50 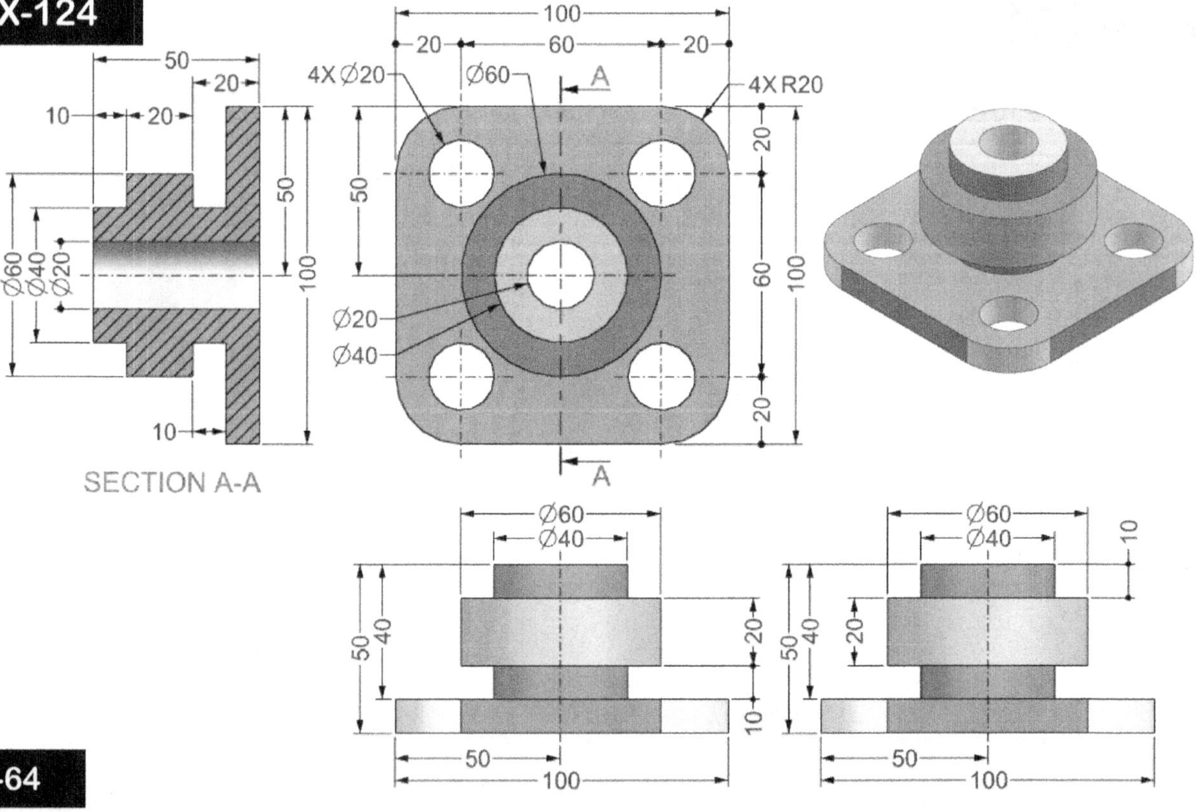
20
10
20
4X Ø20
Ø60
A
4X R20
50
100
20
20
60
100
Ø60
Ø40
Ø20
Ø20
Ø40
A
SECTION A-A
Ø60
Ø40
50
40
20
10
50
100
Ø60
Ø40
50
40
20
10
50
100
100
20
60
20
P-64

EX-125

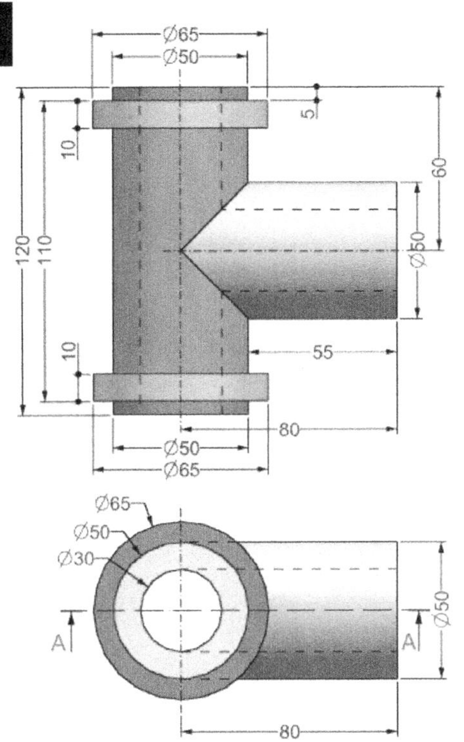
Ø65
Ø50
5
10
60
Ø50
120
110
10
55
10
80
Ø50
Ø65
Ø65
Ø50
Ø30
A
A
Ø50
80

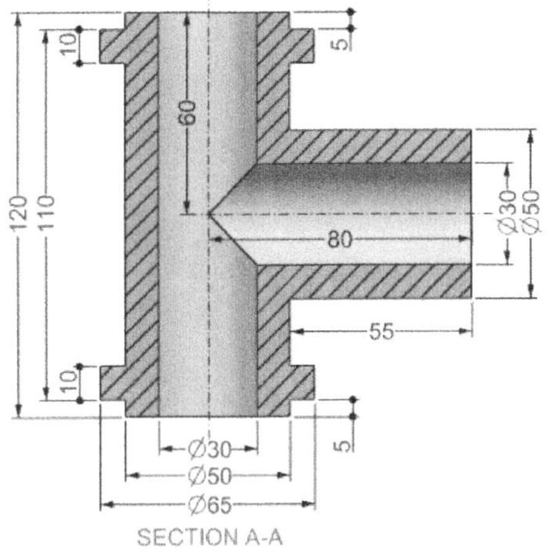
10
5
60
30
50
120
110
80
Ø30
Ø50
10
55
5
Ø30
Ø50
Ø65
SECTION A-A

EX-126

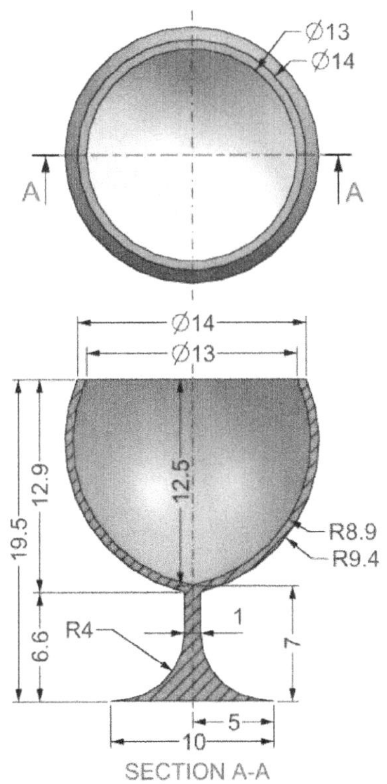
Ø13
Ø14
A
A
Ø14
Ø13
12.9
12.5
19.5
R8.9
R9.4
6.6
R4
1
7
5
10
SECTION A-A

P-65

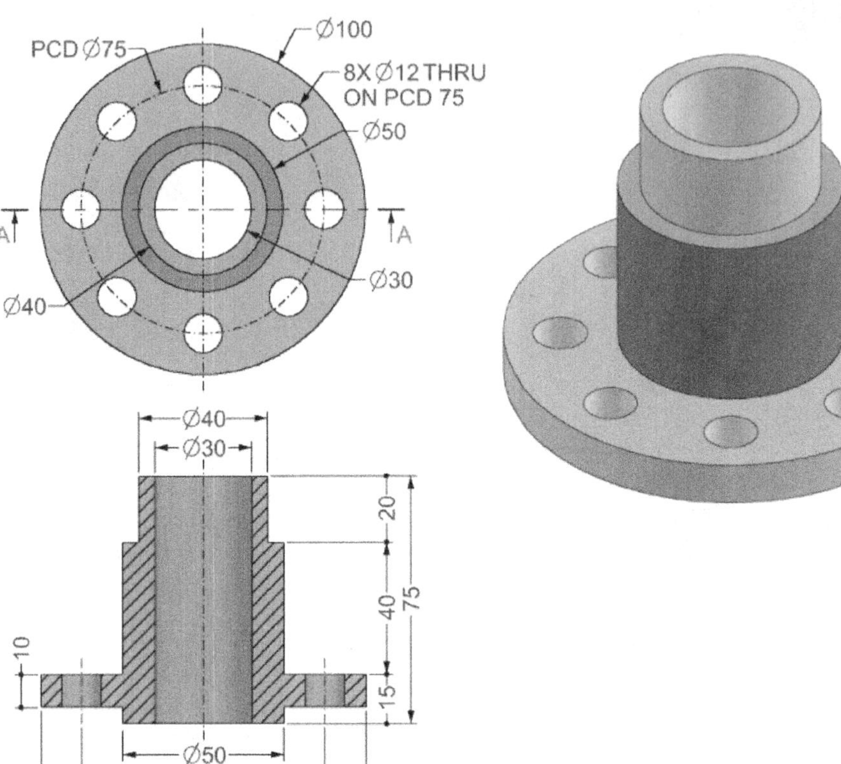

PCD Ø75
Ø100
8X Ø12 THRU ON PCD 75
Ø50
Ø30
A
A
Ø40
Ø40
Ø30
20
40
75
10
15
Ø50
75
Ø100
SECTION A-A

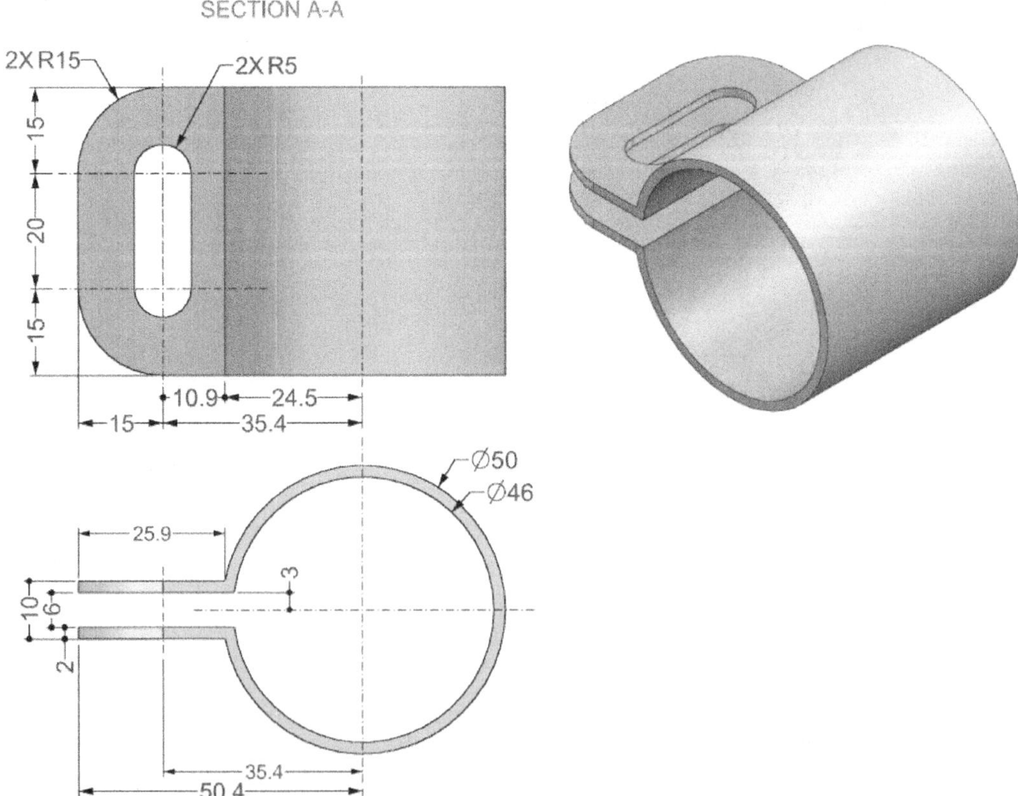

2X R15
2X R5
15
20
15
10.9
24.5
15
35.4
Ø50
Ø46
25.9
3
10
6
2
35.4
50.4

EX-129

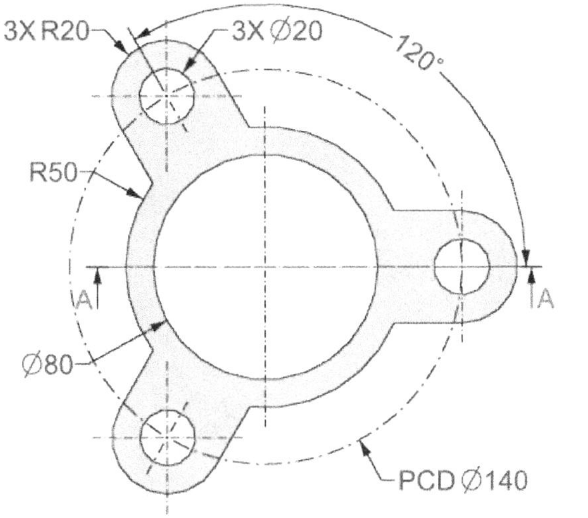

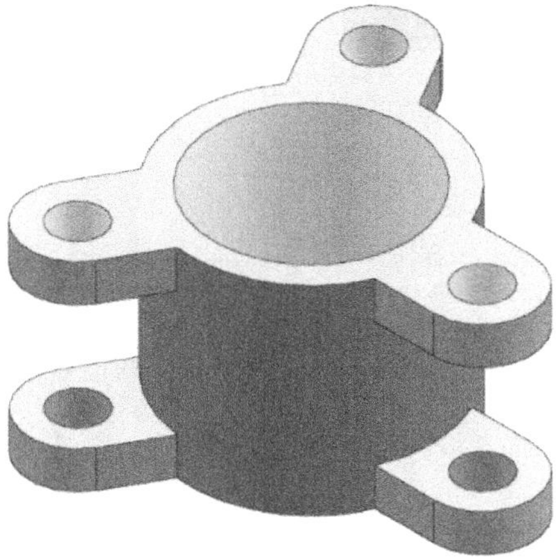

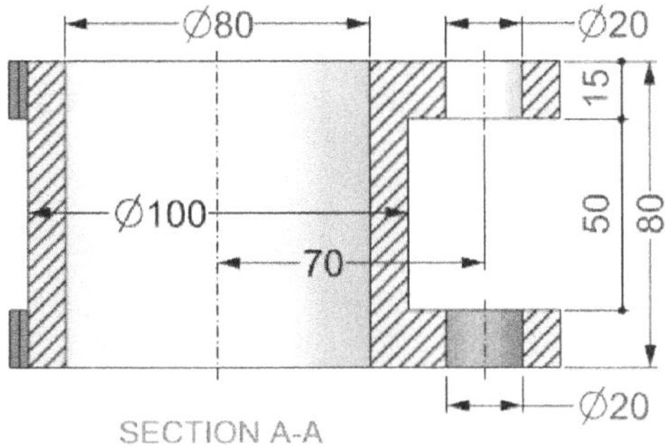

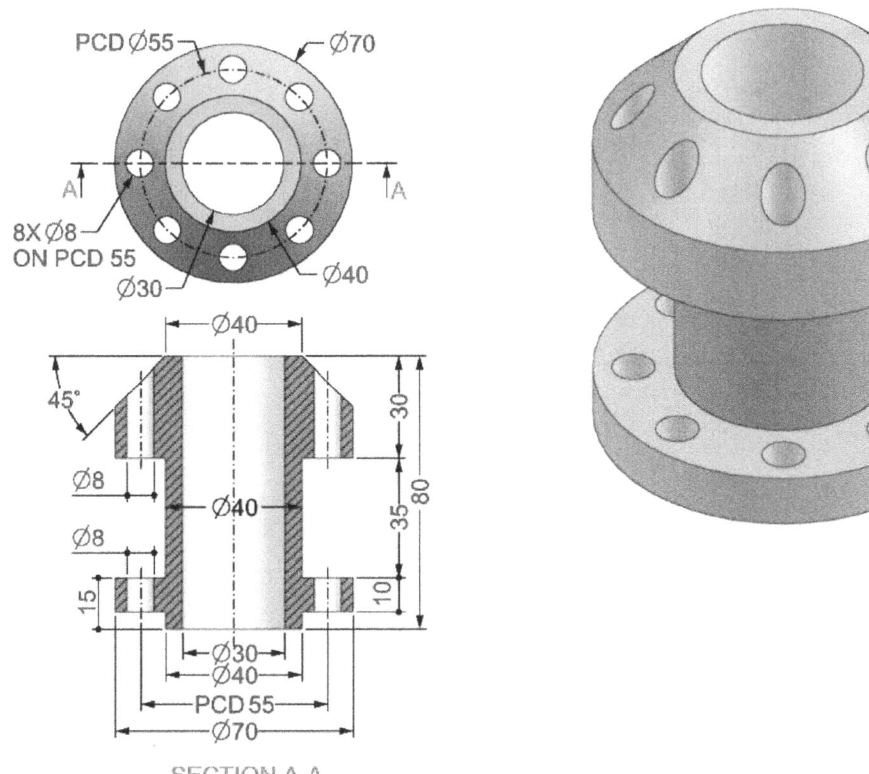
3X R20
3X Ø20
120°
R50
A
A
Ø80
PCD Ø140
Ø80
Ø20
15
Ø100
50
80
70
Ø20
SECTION A-A
EX-130
PCD Ø55
Ø70
A
A
8X Ø8
ON PCD 55
Ø30
Ø40
Ø40
45°
30
Ø8
Ø40
80
35
Ø8
10
15
Ø30
Ø40
PCD 55
Ø70
SECTION A-A

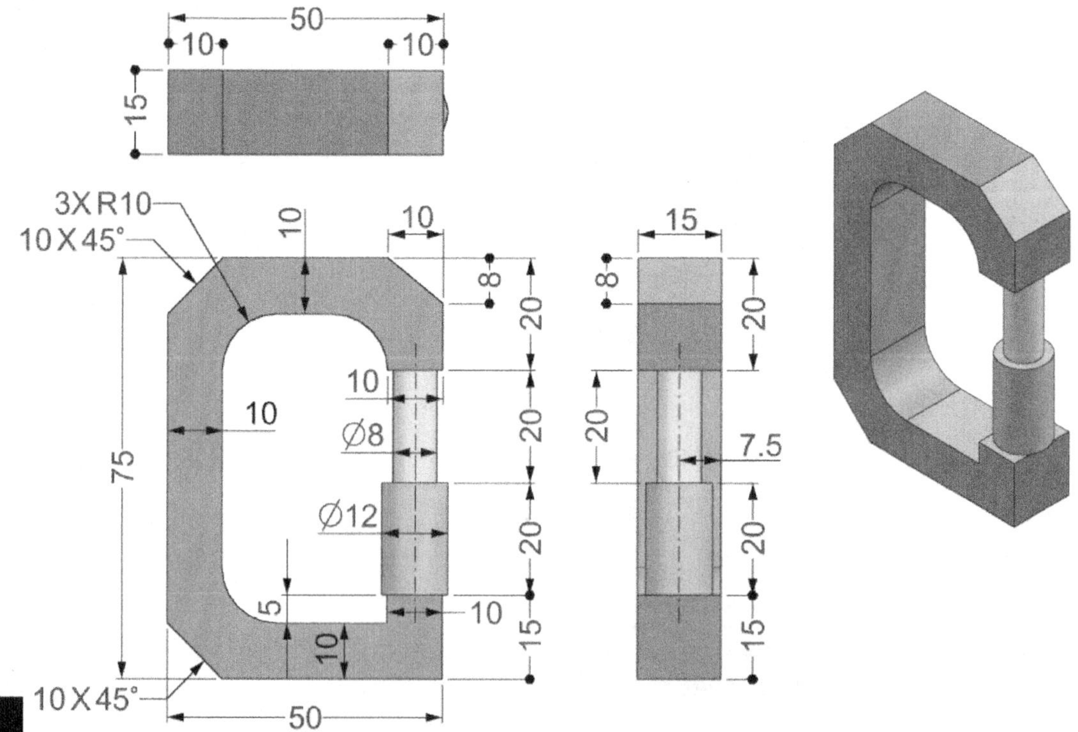
3X R20
3X Ø20
2X R60
94.2
54.2
80
45.8
40
20
40
50
50
20
50
20
20
20
50
50
94.2
R30
Ø30
R15
20
10
40
80
60
50
10
10
15
3X R10
10 X 45°
10
10
10
8
20
20
20
10
Ø8
Ø12
5
10
75
50
15
20
20
20
15
8
8
20
20
7.5
20
15

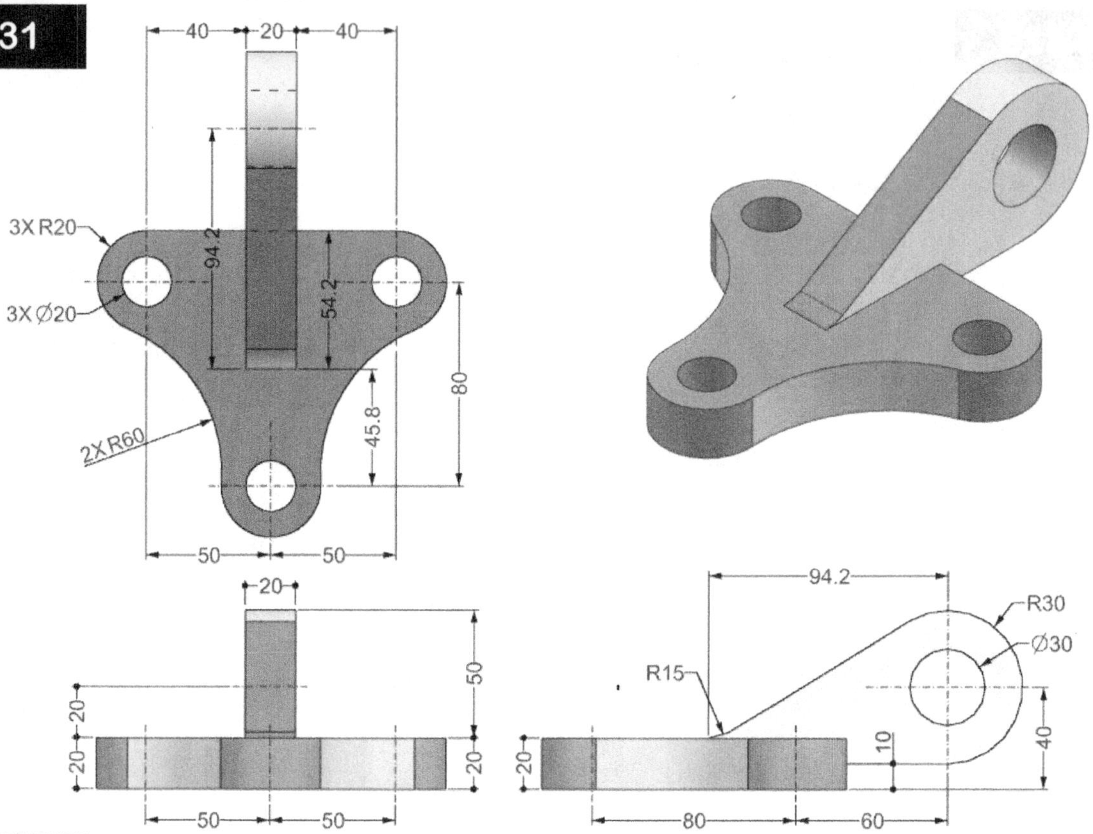

EX-133 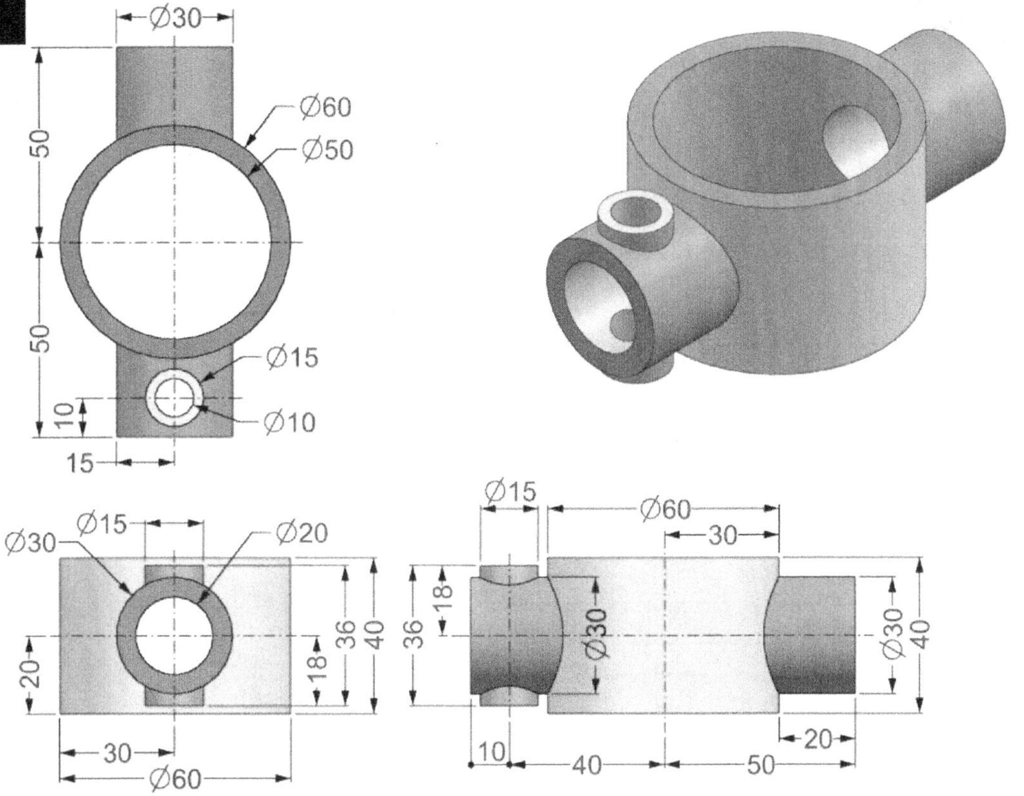
Ø30
Ø60
Ø50
50
50
Ø15
Ø10
10
15
Ø15
Ø30
Ø20
36
40
20
18
30
Ø60
Ø15
Ø60
30
18
36
Ø30
Ø30
40
10
40
50
20
EX-134 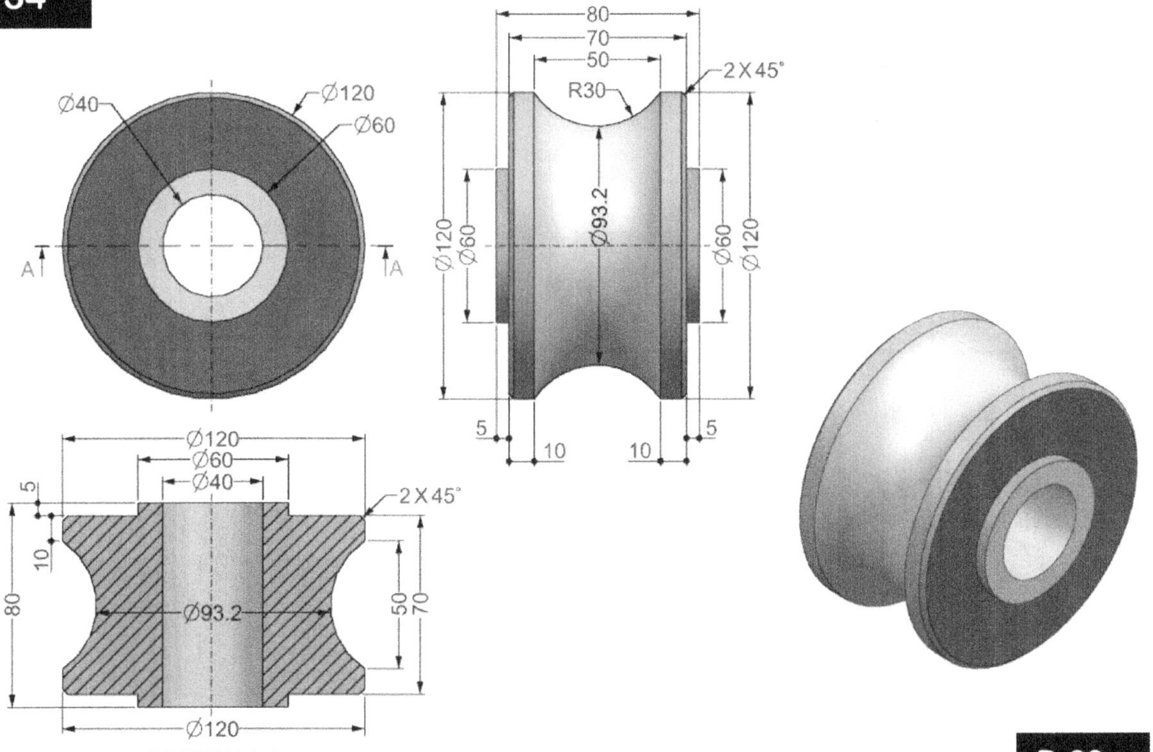
Ø40
Ø120
Ø60
A
A
80
70
50
R30
2 X 45°
Ø120
Ø60
Ø93.2
Ø60
Ø120
5
10
10
5
Ø120
Ø60
Ø40
2 X 45°
5
10
80
Ø93.2
50
70
Ø120
SECTION A-A
P-69

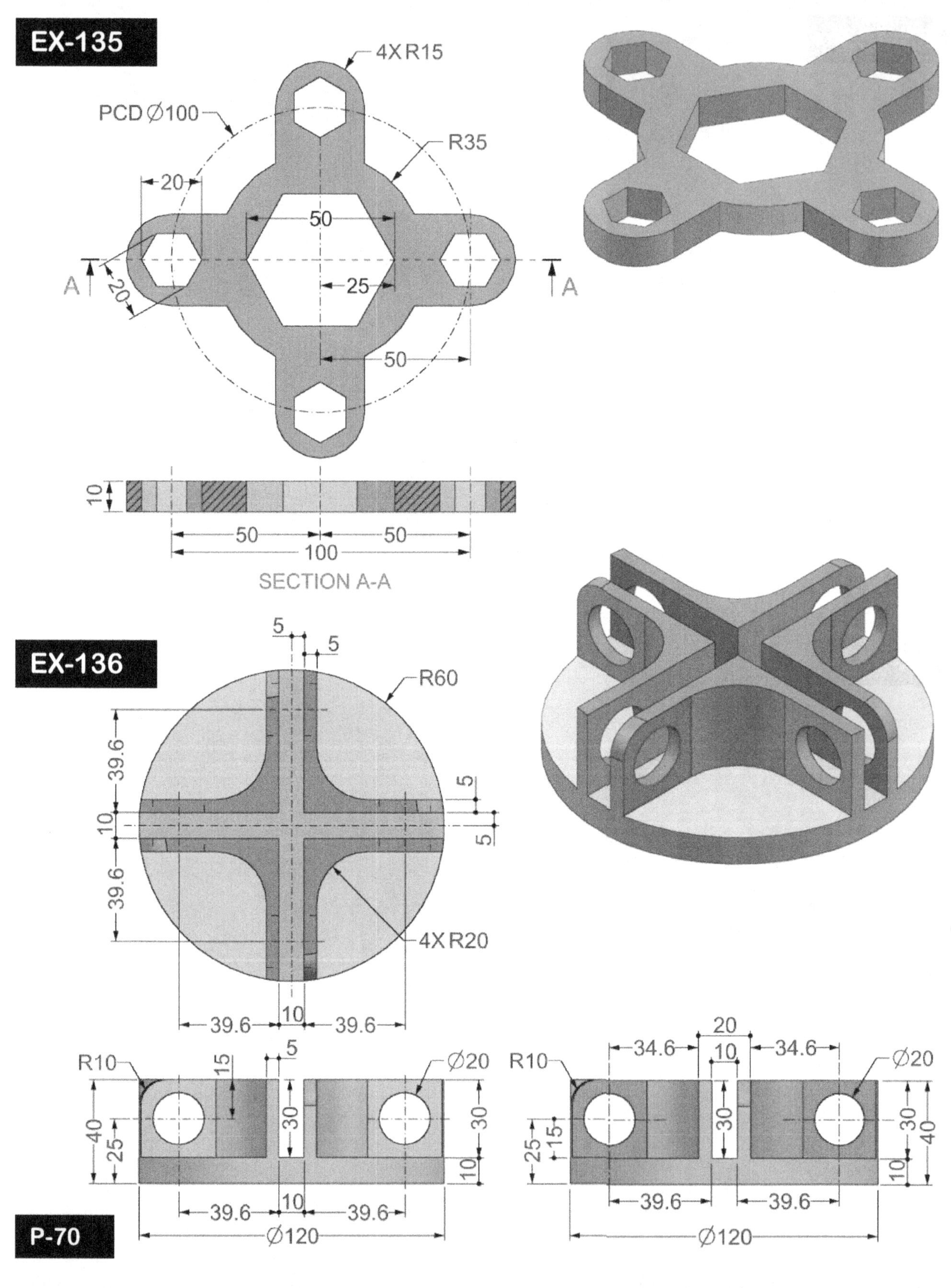

EX-135
4X R15
PCD Ø100
R35
20
50
25
20
50
A
A
10
50
50
100
SECTION A-A
EX-136
5
5
R60
39.6
39.6
10
5
5
39.6
39.6
4X R20
10
39.6
10
39.6
5
R10
15
5
Ø20
40
25
30
30
10
39.6
10
39.6
Ø120
20
34.6
10
34.6
R10
Ø20
25
15
30
30
40
10
39.6
39.6
Ø120
P-70

EX-137
2X R6
2X R5
20
A
A
B
6
10
10
15
5
10
10
10
20
10
34.6
10.2
55
3
SECTION A-A
(SCALE 1:1)
R1
1
1
1
1
DETAIL B
(SCALE 2:1)
SHELL THICKNESS = 1MM
ALL INSIDE WALL THICKNESS

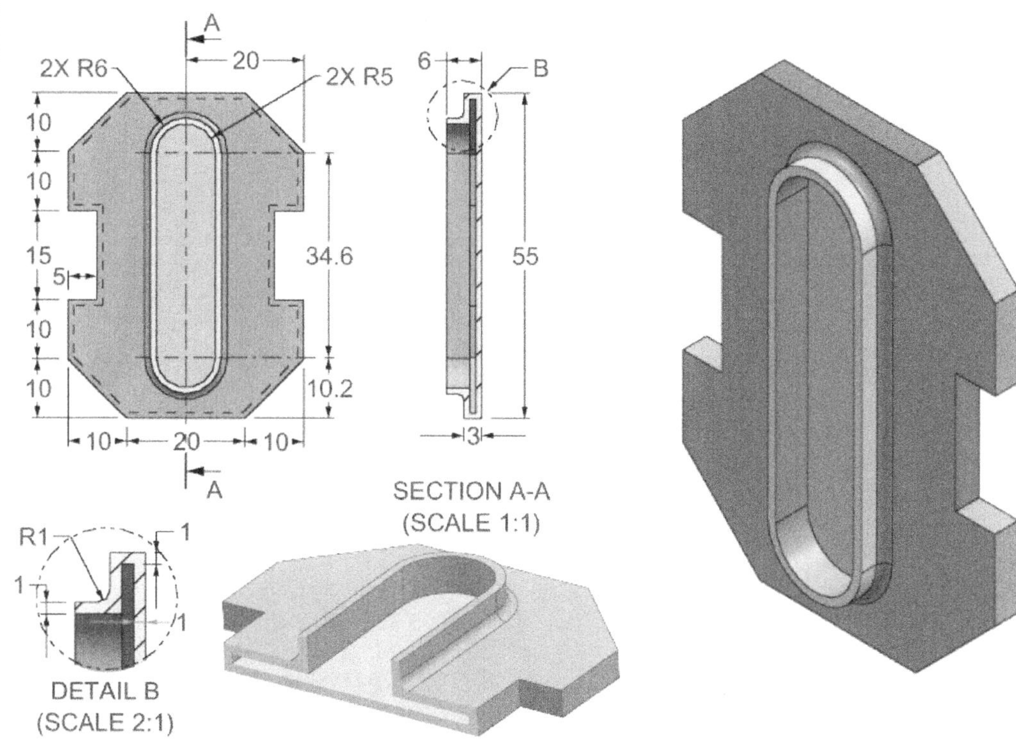

EX-138
60°
60°
R46
R41
R50
R37
60°
9
60°
Ø12
Ø12
Ø12
60°
Ø12
10
20

P-71

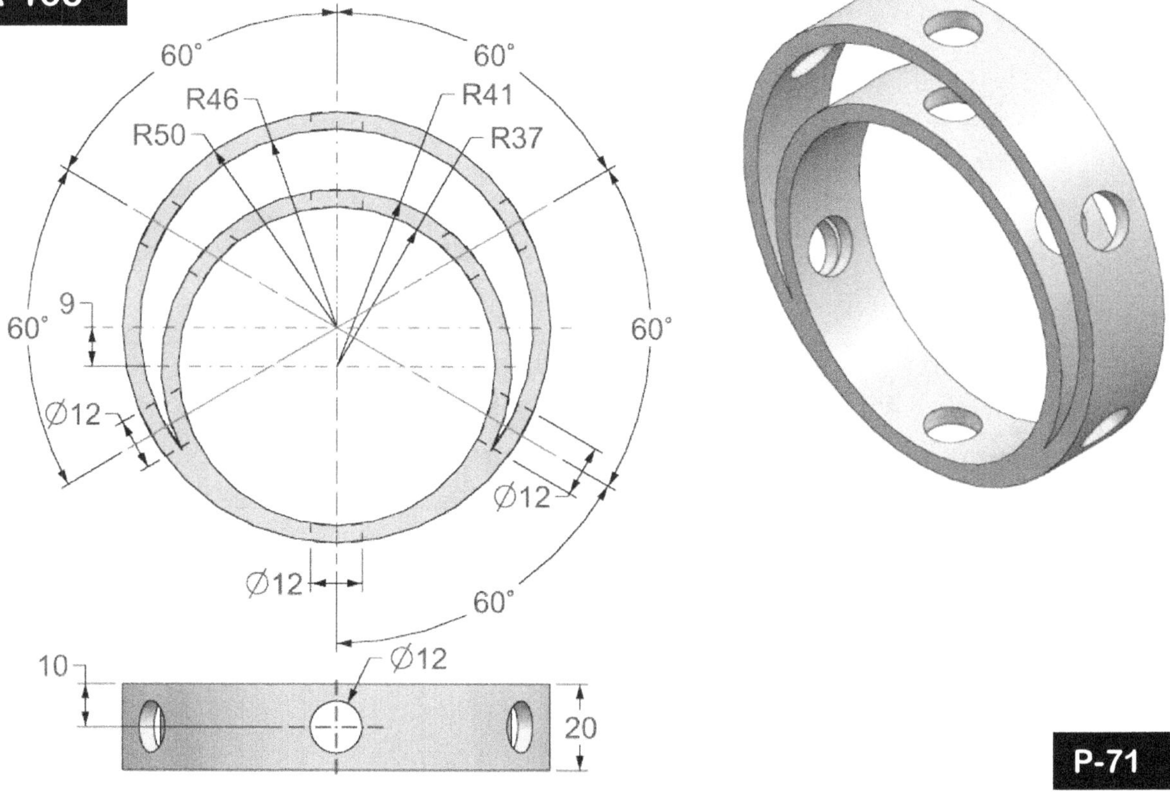

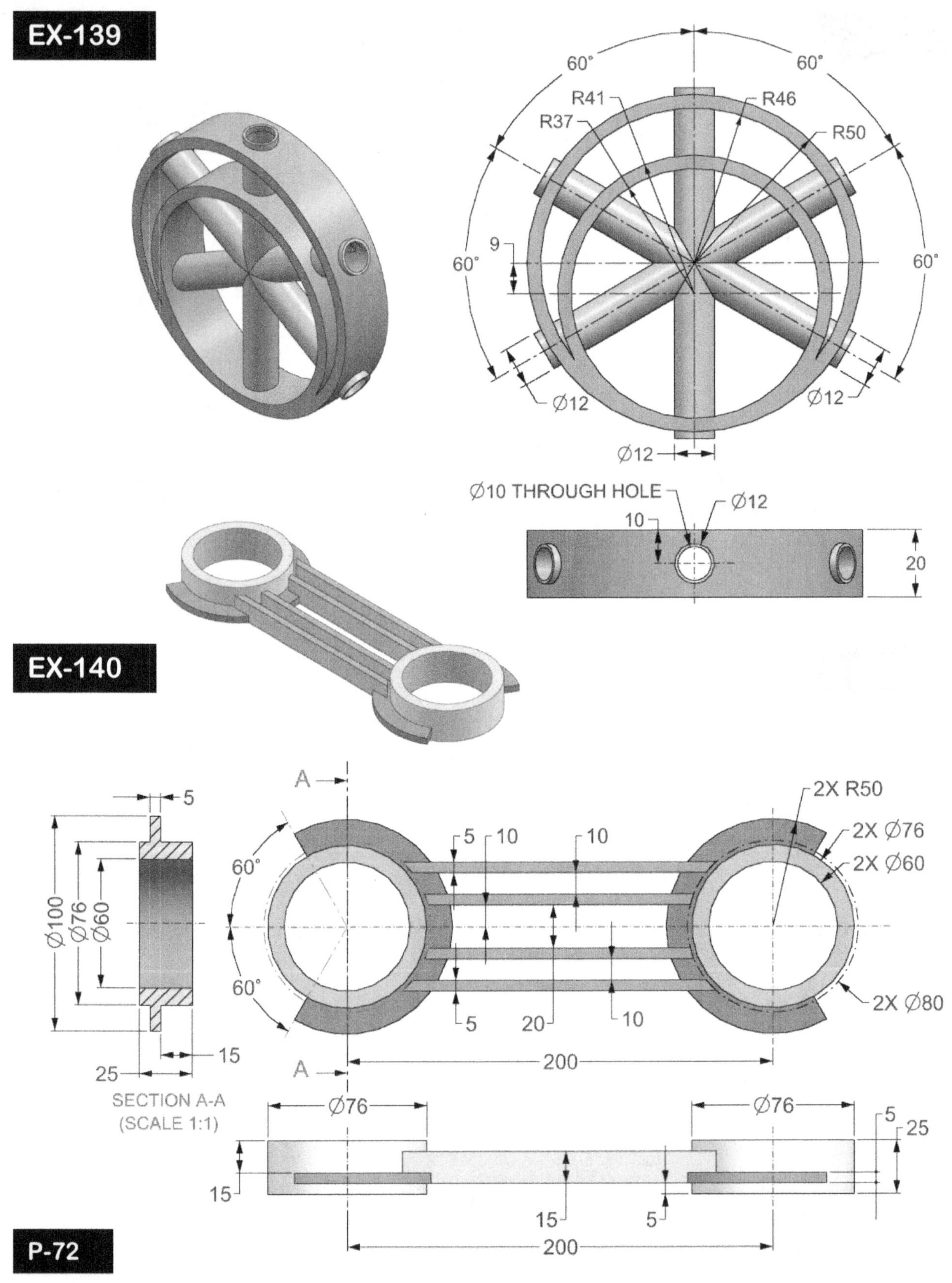

EX-139
60°
60°
R41
R46
R37
R50
9
60°
60°
60°
60°
Ø12
Ø12
Ø12
Ø10 THROUGH HOLE
Ø12
10
20
EX-140
A
5
2X R50
2X Ø76
2X Ø60
5
10
10
60°
Ø100
Ø76
Ø60
5
10
60°
2X Ø80
20
10
A
15
200
25
SECTION A-A
(SCALE 1:1)
Ø76
Ø76
5
25
15
15
5
15
200
P-72

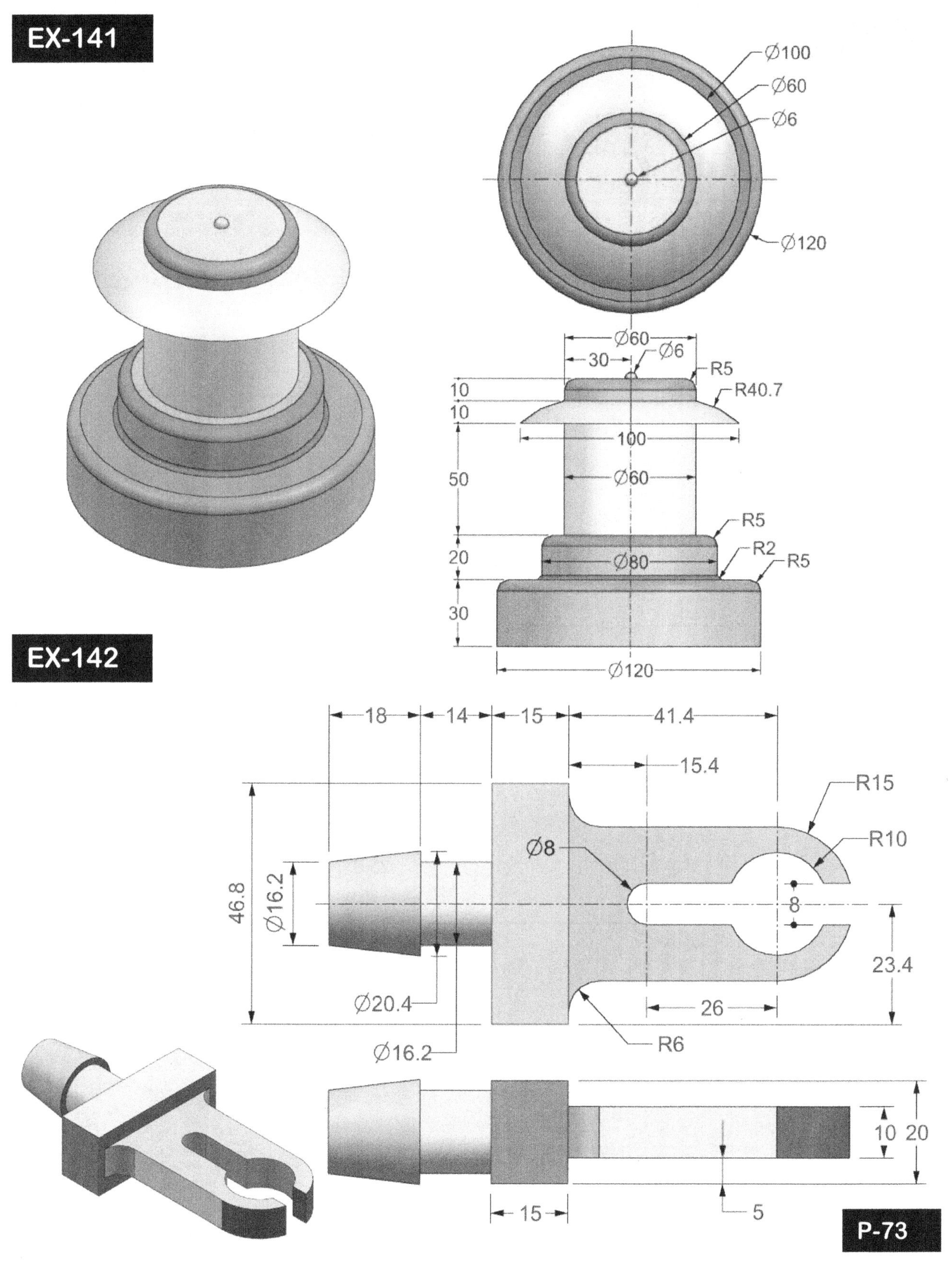

EX-141
EX-142
Ø100
Ø60
Ø6
Ø120
Ø60
30
Ø6
R5
R40.7
10
10
100
Ø60
50
R5
20
R2
R5
Ø80
30
Ø120
18
14
15
41.4
15.4
R15
Ø8
R10
46.8
Ø16.2
8
Ø20.4
23.4
26
Ø16.2
R6
10 20
15
5
P-73

EX-143

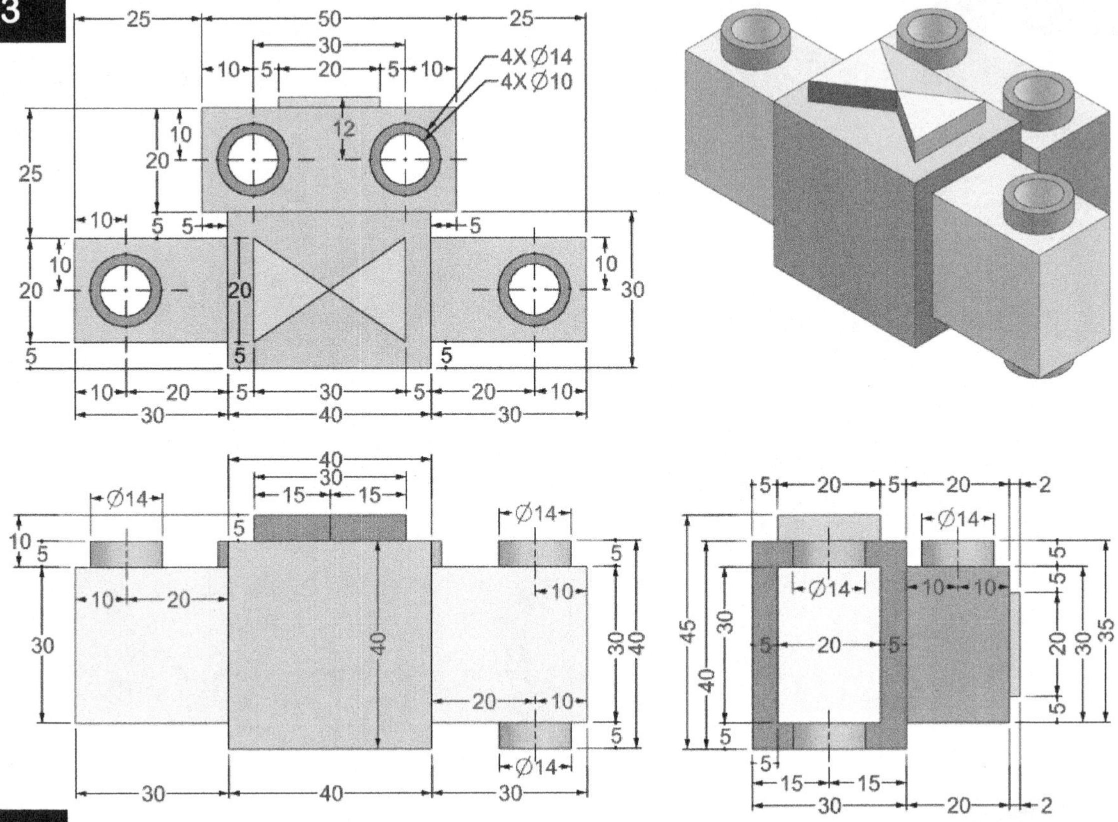

EX-144

4X Ø14
4X Ø10

12

Ø14

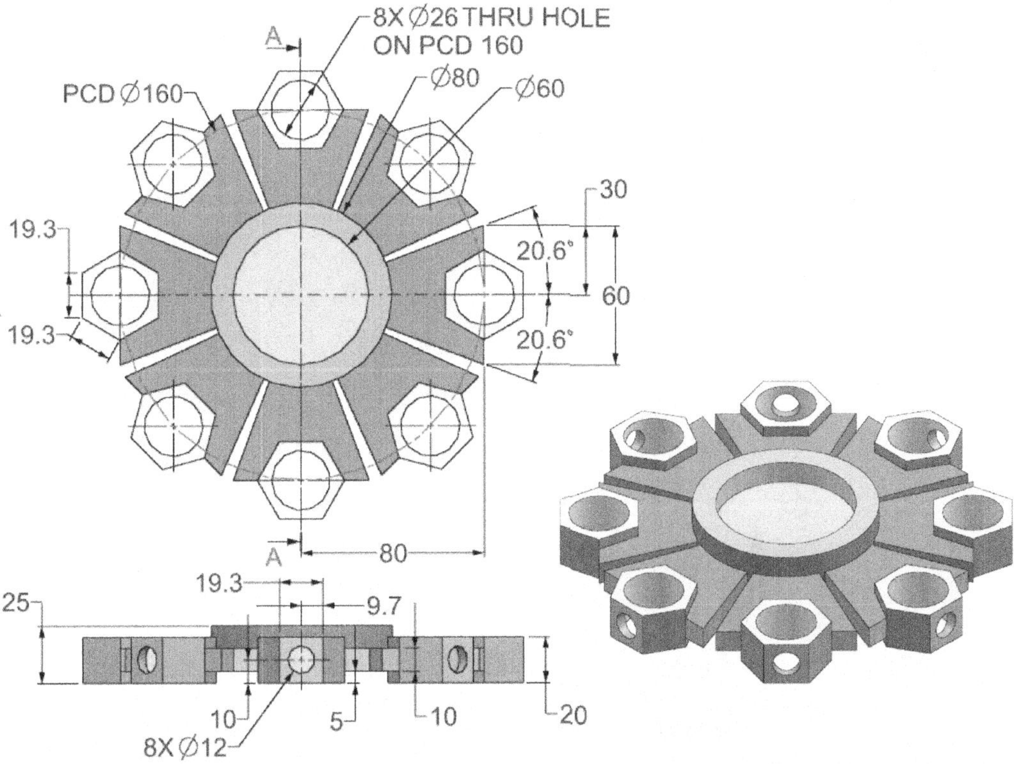

8X Ø26 THRU HOLE
ON PCD 160

PCD Ø160

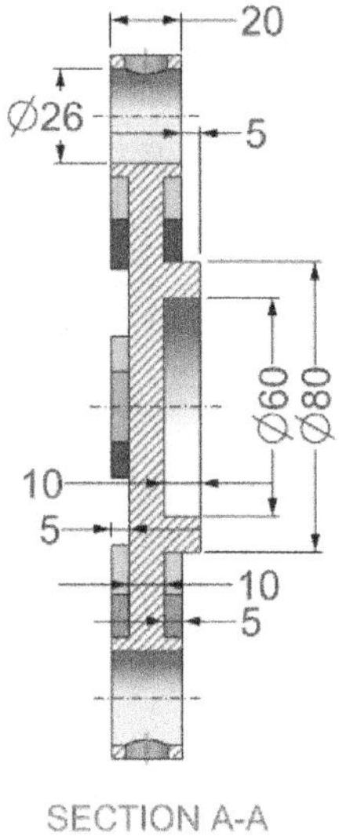

Ø80

Ø60

SECTION A-A

19.3

9.7

8X Ø12

P-74

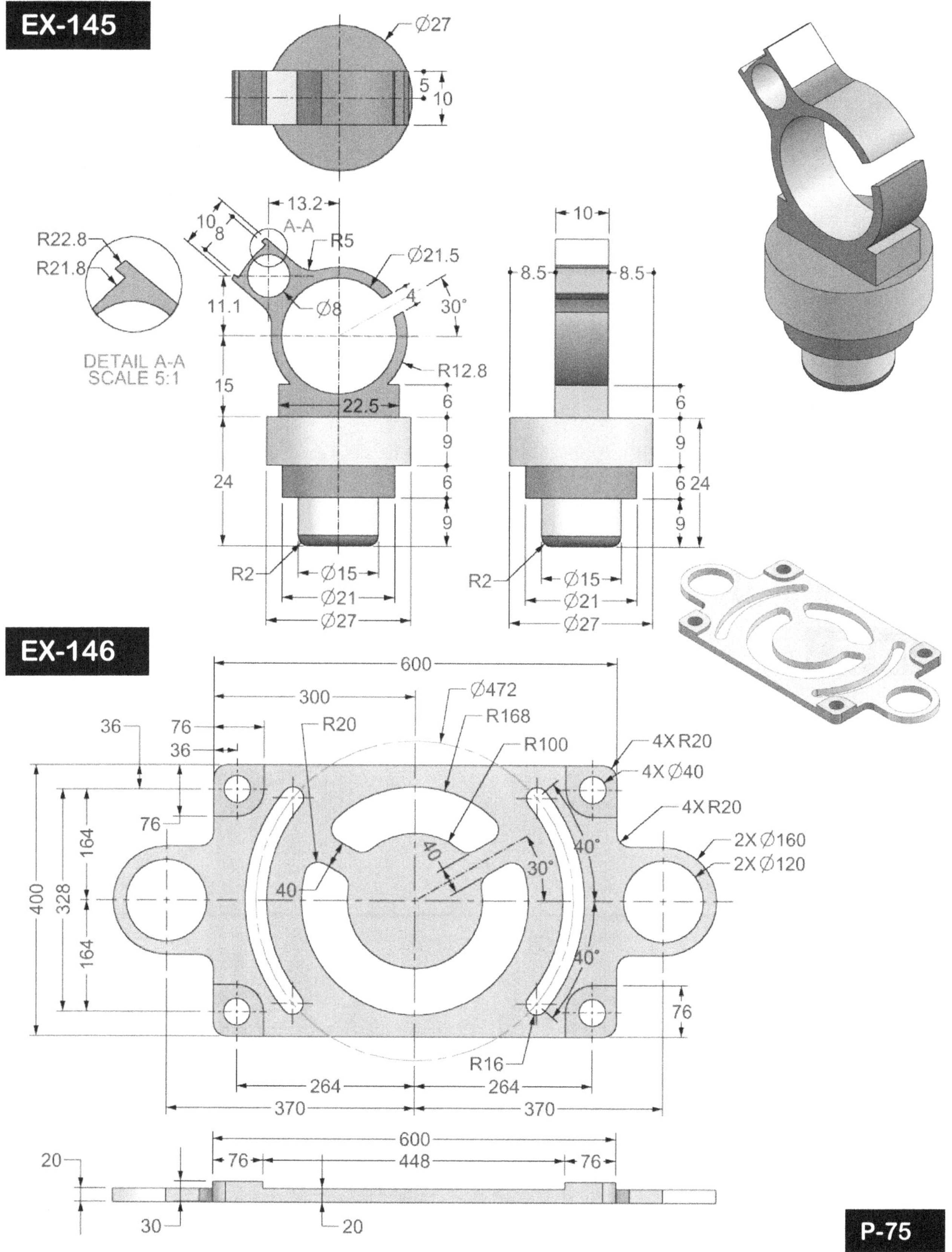

EX-145
Ø27
5
10
13.2
10
8
A-A
R5
Ø21.5
R22.8
R21.8
4
30°
11.1
Ø8
DETAIL A-A
SCALE 5:1
R12.8
15
22.5
6
9
6
24
9
R2
Ø15
Ø21
Ø27
10
8.5
8.5
6
9
6 24
9
R2
Ø15
Ø21
Ø27
EX-146
600
300
Ø472
R168
R20
R100
4X R20
4X Ø40
36
76
36
4X R20
76
2X Ø160
2X Ø120
164
40
40
30°
40°
400
328
40
164
40°
76
R16
264
264
370
370
600
76
448
76
20
30
20
P-75

EX-147

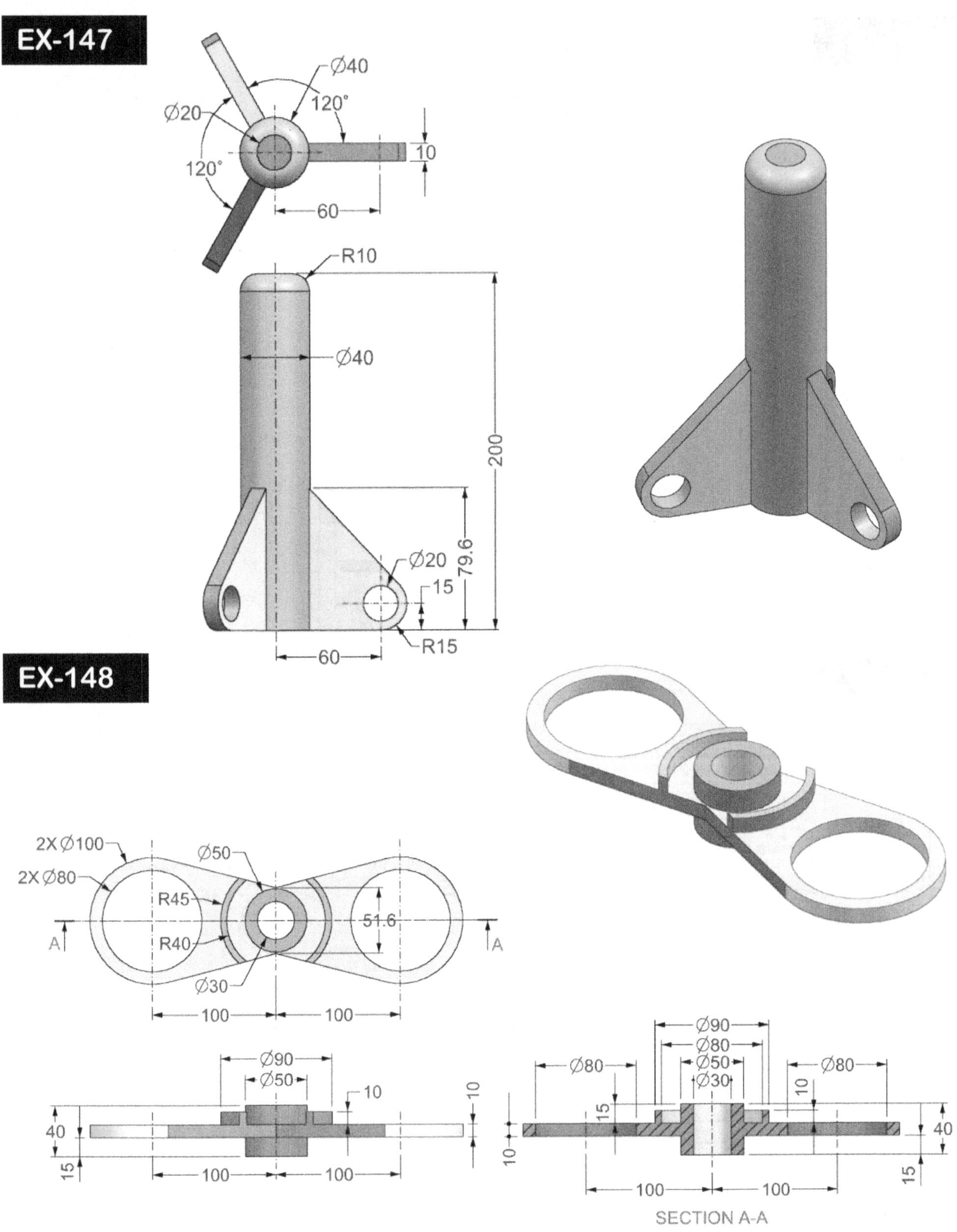
Ø40
Ø20
120°
120°
10
60
R10
Ø40
200
79.6
Ø20
15
60
R15
EX-148
2X Ø100
2X Ø80
Ø50
R45
R40
51.6
Ø30
A
A
100
100
Ø90
Ø50
10
10
40
15
100
100
Ø90
Ø80
Ø50
Ø30
Ø80
Ø80
15
15
10
10
40
100
100
SECTION A-A

P-76

EX-149
Ø24
2X R5
R24
R30
Ø41.7
18
Ø15
22.2
35.9
100
R26.9
A
A
100
15
5
Ø15
Ø24
5
Ø41.7
SECTION A-A
EX-150
67.5
17.8
R16.7
R21.5
13.3
14.4
10
R9.6
R4
Ø12
R15.7
R6.1
R6.2
A
A
2X R19.2
45.9
3.6
19
5
Ø12
SECTION A-A
P-77

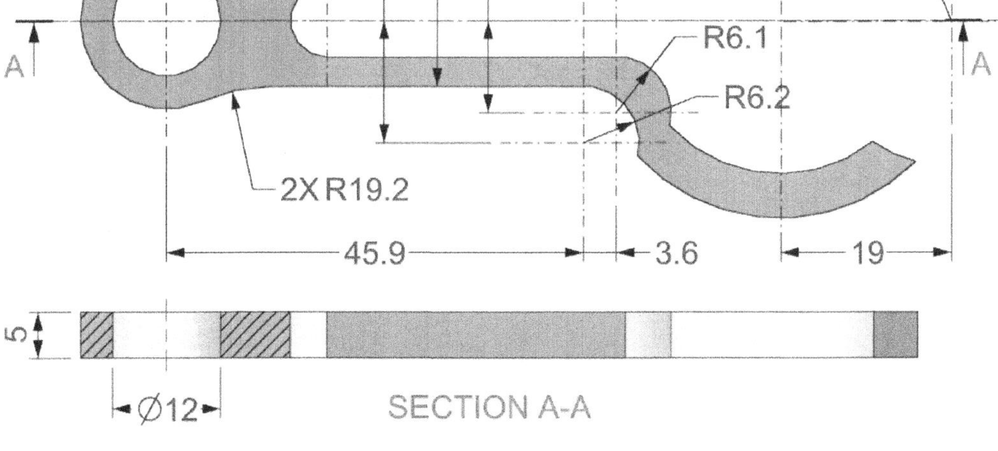

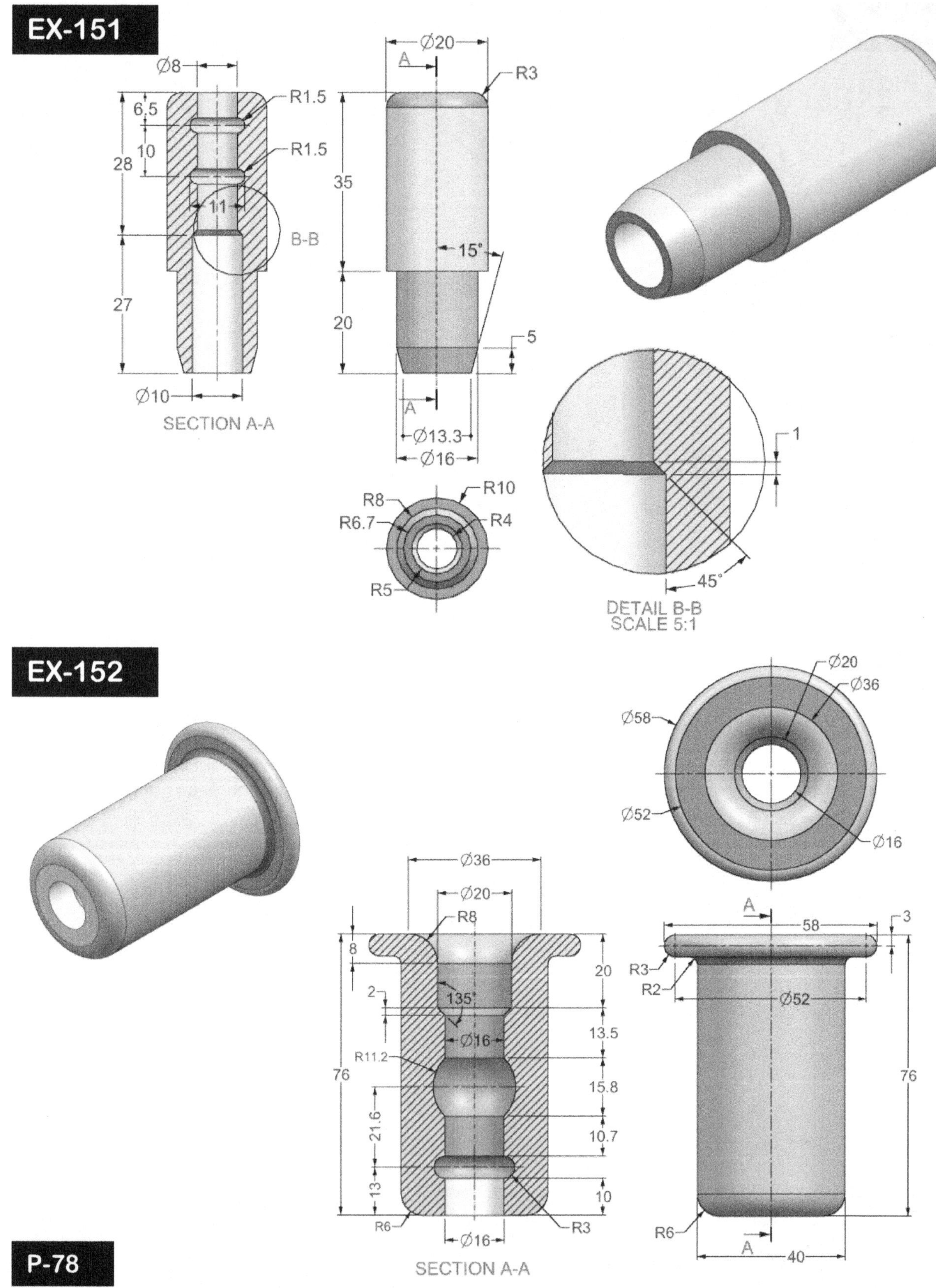

EX-151
SECTION A-A
Ø8
Ø10
6.5
10
28
27
R1.5
R1.5
1.1
B-B
Ø20
A
R3
35
20
15°
5
A
Ø13.3
Ø16
R8
R10
R6.7
R4
R5
DETAIL B-B
SCALE 5:1
45°
1
EX-152
Ø20
Ø36
Ø58
Ø52
Ø16
SECTION A-A
Ø36
Ø20
R8
8
2
135°
Ø16
R11.2
76
21.6
13
R6
Ø16
R3
20
13.5
15.8
10.7
10
A
58
3
R3
R2
Ø52
76
R6
A
40
P-78

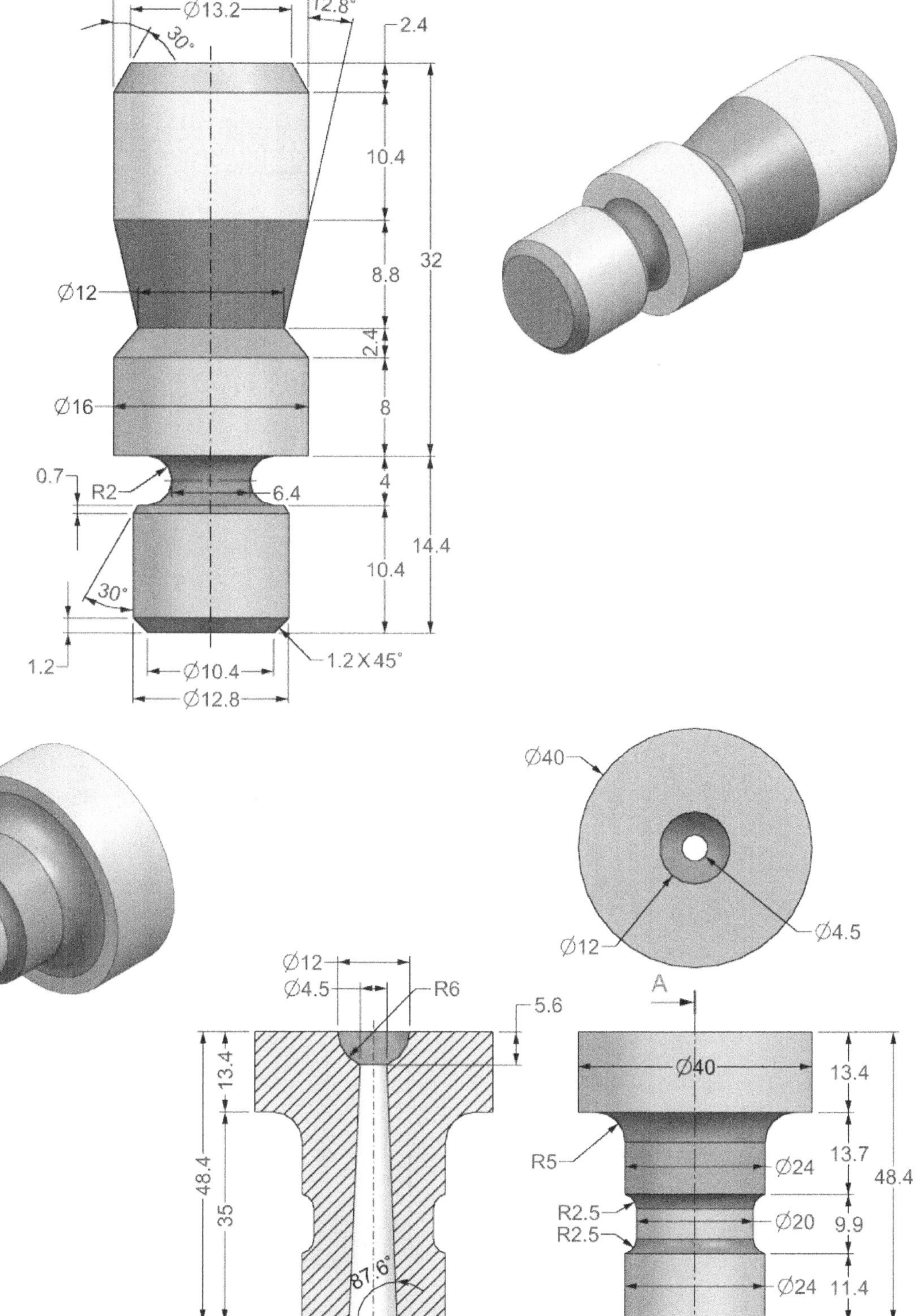

EX-153
EX-154
P-79
SECTION A-A
Ø16
Ø13.2
12.8°
30°
2.4
10.4
32
8.8
Ø12
2.4
Ø16
8
0.7
R2
6.4
4
14.4
10.4
30°
1.2
Ø10.4
Ø12.8
1.2 X 45°
Ø40
Ø12
Ø4.5
A
Ø12
Ø4.5
R6
5.6
13.4
48.4
35
87.6°
Ø8
Ø40
13.4
R5
13.7
R2.5
Ø24
R2.5
Ø20
9.9
48.4
Ø24
11.4
A

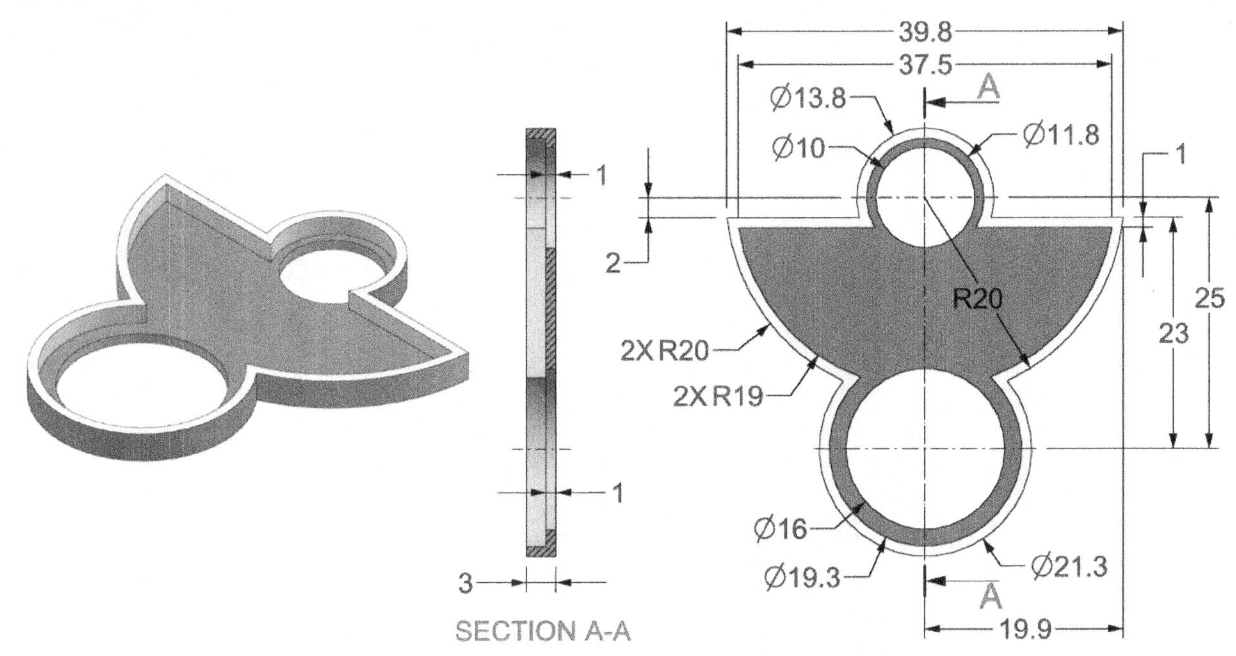

EX-155

39.8
37.5
Ø13.8
Ø10
Ø11.8
A
1
2
1
R20
25
23
2X R20
2X R19
Ø16
Ø19.3
Ø21.3
A
19.9
3
SECTION A-A

EX-156

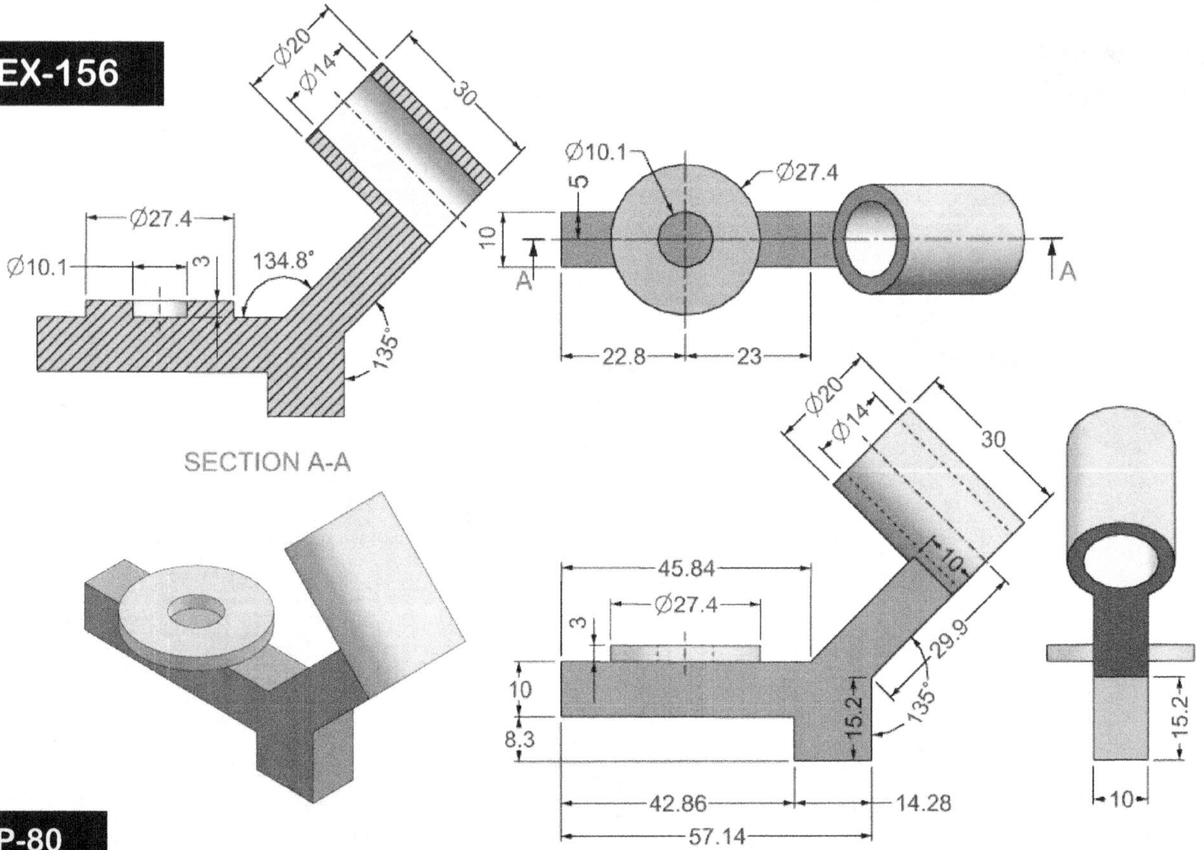

Ø20
Ø14
30
Ø27.4
Ø10.1
3
134.8°
135°
SECTION A-A

Ø10.1
Ø27.4
5
10
A
22.8
23
A

Ø20
Ø14
30
10
45.84
Ø27.4
3
10
8.3
42.86
14.28
57.14
15.2
29.9
135°
15.2
10

P-80

EX-157

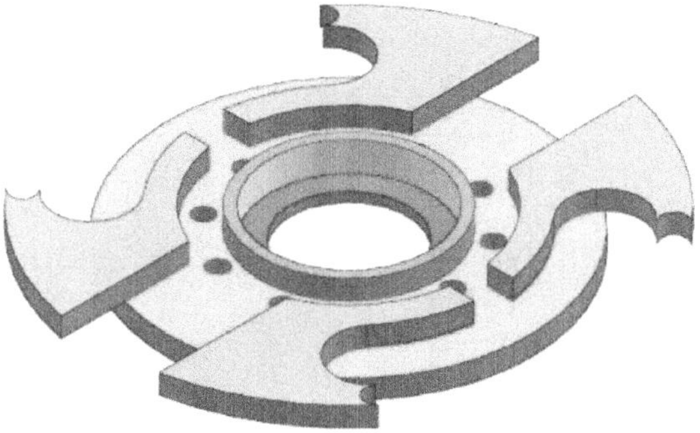

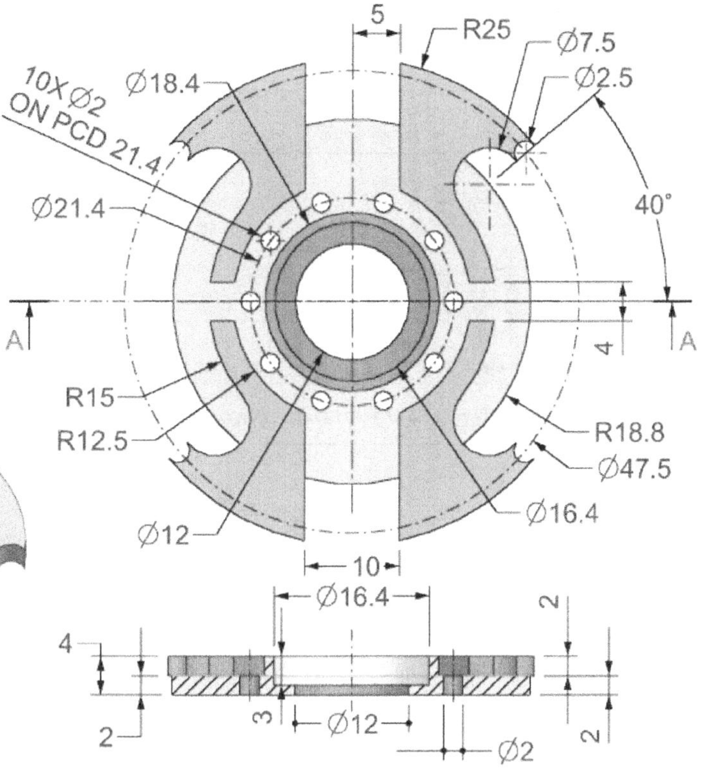

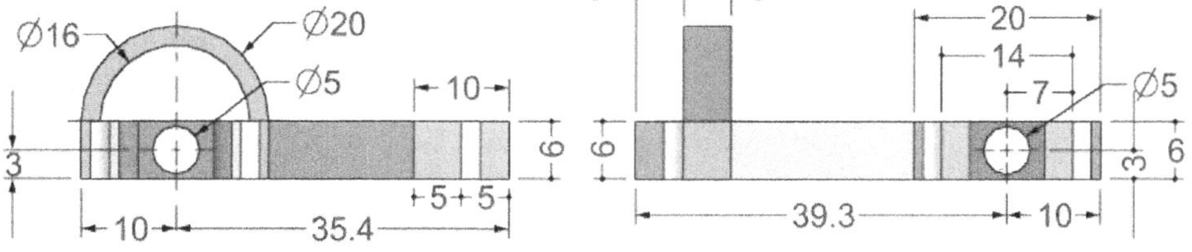

EX-158

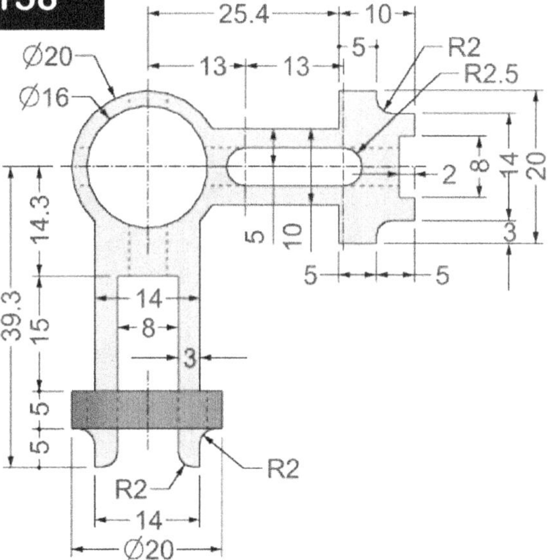

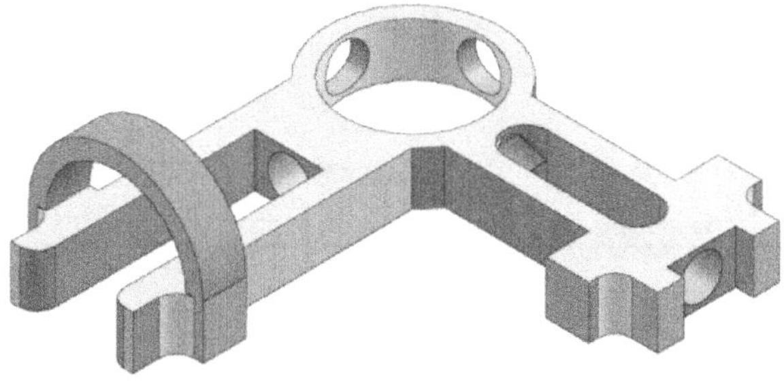

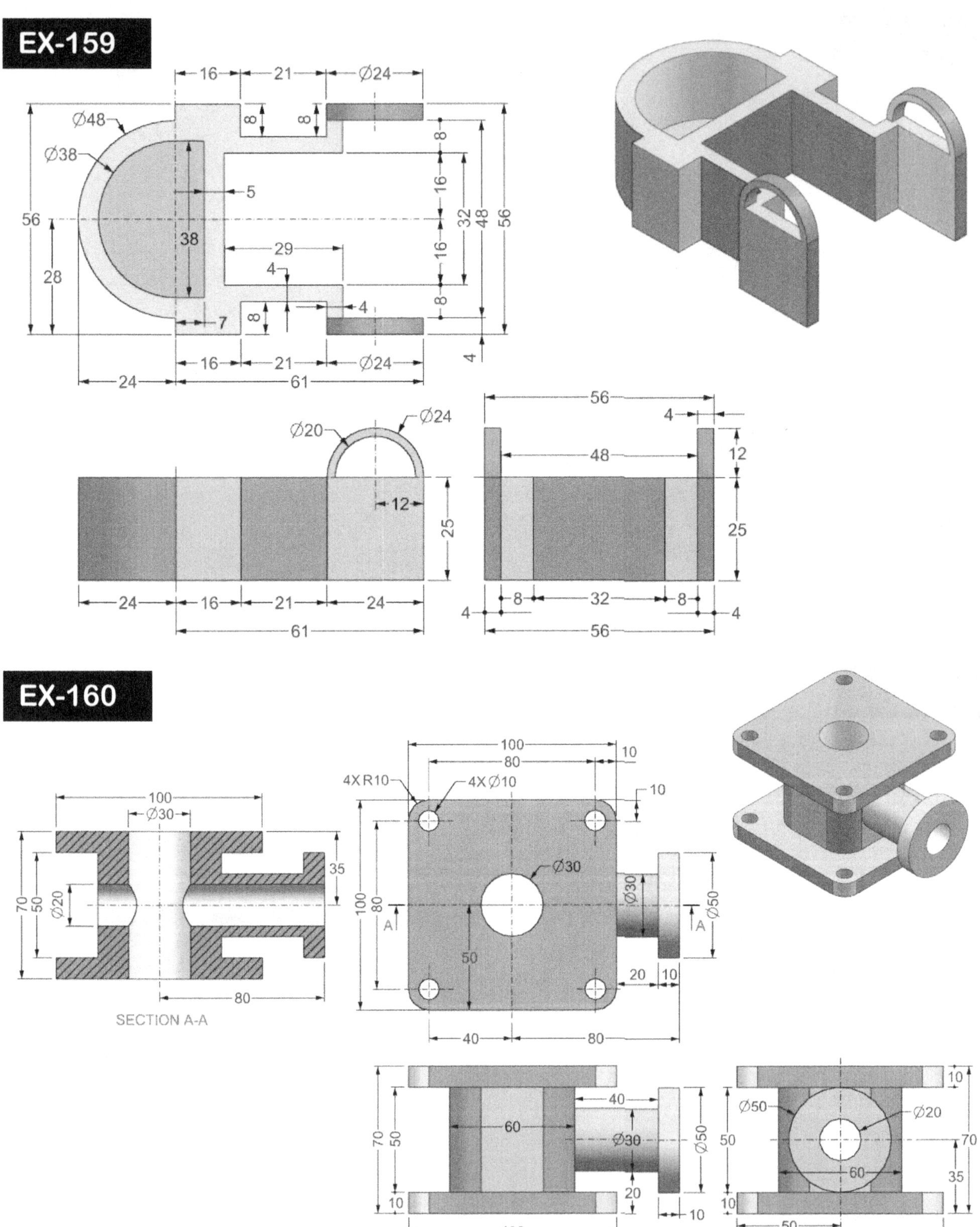

Ø48
Ø38
56
28
24
16
21
Ø24
8
8
5
38
29
4
7
8
16
61
8
16
32
48
16
8
56
4
Ø24
4
Ø20
Ø24
12
25
24
16
21
24
61
56
4
48
12
25
8
32
8
4
56
4

100
Ø30
35
70
50
Ø20
80
SECTION A-A
100
80
10
4X R10
4X Ø10
10
Ø30
Ø30
Ø50
A
A
50
40
80
20
10
70
50
60
40
Ø30
Ø50
20
10
100
10
Ø50
Ø20
50
60
70
35
10
50
100

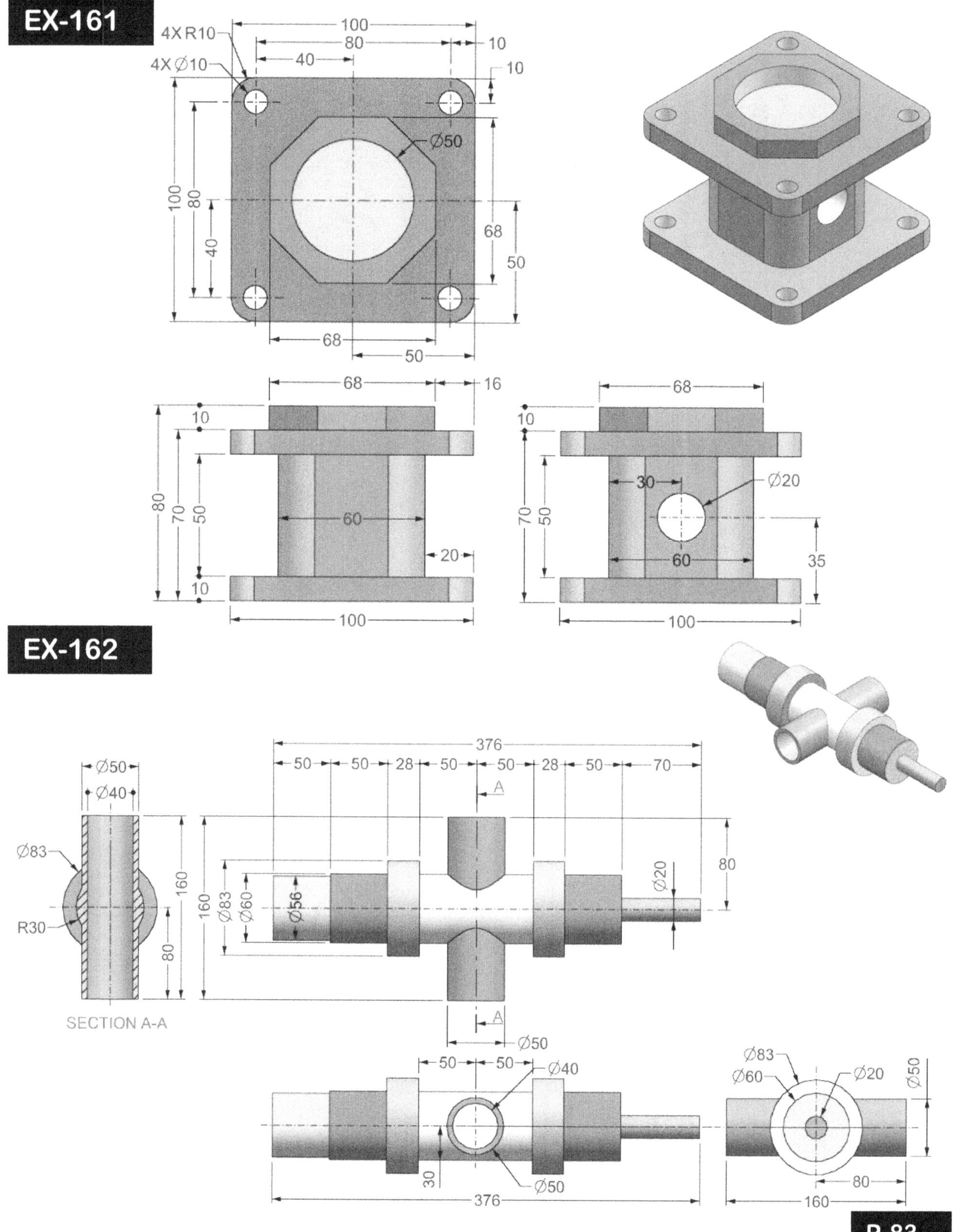

EX-161
4X R10
4X Ø10
100
80
40
10
10
Ø50
100
80
40
68
50
68
50
68
16
10
80
70
50
60
20
10
100
68
10
70
50
30
Ø20
60
35
100
EX-162
376
50
50
28
50
50
28
50
70
A
Ø50
Ø40
Ø83
R30
160
160
80
Ø83
Ø60
Ø56
Ø20
80
SECTION A-A
A
Ø50
50
50
Ø40
30
Ø50
376
Ø83
Ø60
Ø20
Ø50
80
160
P-83

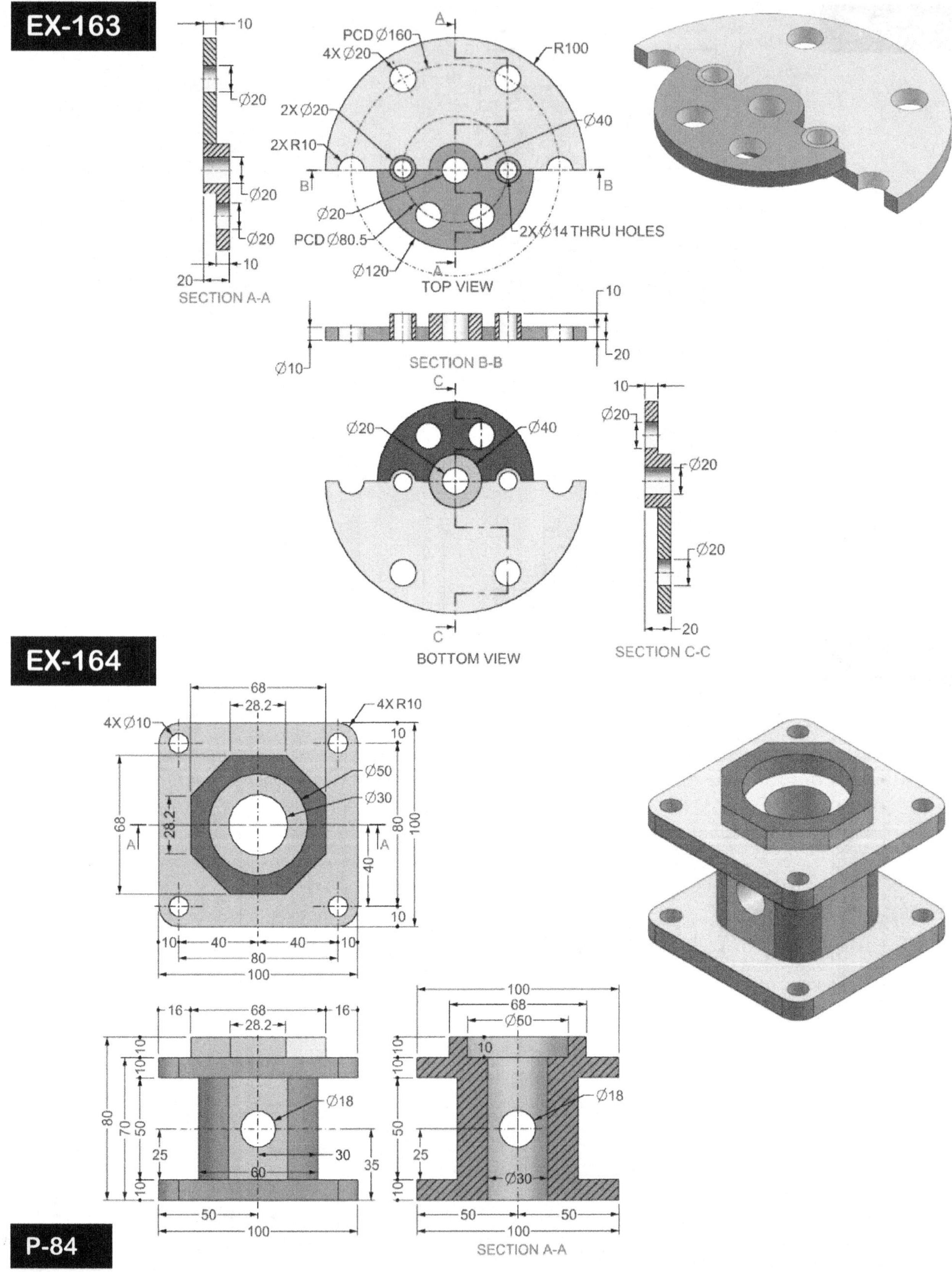

EX-163
10
PCD Ø160
4X Ø20
2X Ø20
2X R10
Ø20
PCD Ø80.5
Ø120
Ø20
Ø20
Ø20
10
20
SECTION A-A
A
R100
Ø40
B
B
Ø20
2X Ø14 THRU HOLES
TOP VIEW
10
20
Ø10
SECTION B-B
C
Ø20
Ø40
Ø20
BOTTOM VIEW
C
10
Ø20
Ø20
Ø20
20
SECTION C-C
EX-164
68
28.2
4X Ø10
4X R10
10
Ø50
Ø30
68
28.2
80
100
40
A
A
10
10
40
40
10
80
100
16
68
16
28.2
10 10
Ø18
80
70
50
25
30
60
10
50
100
100
68
Ø50
10 10
10
Ø18
50
25
Ø30
10
35
50
50
100
SECTION A-A
P-84

EX-165
EX-166
P-85

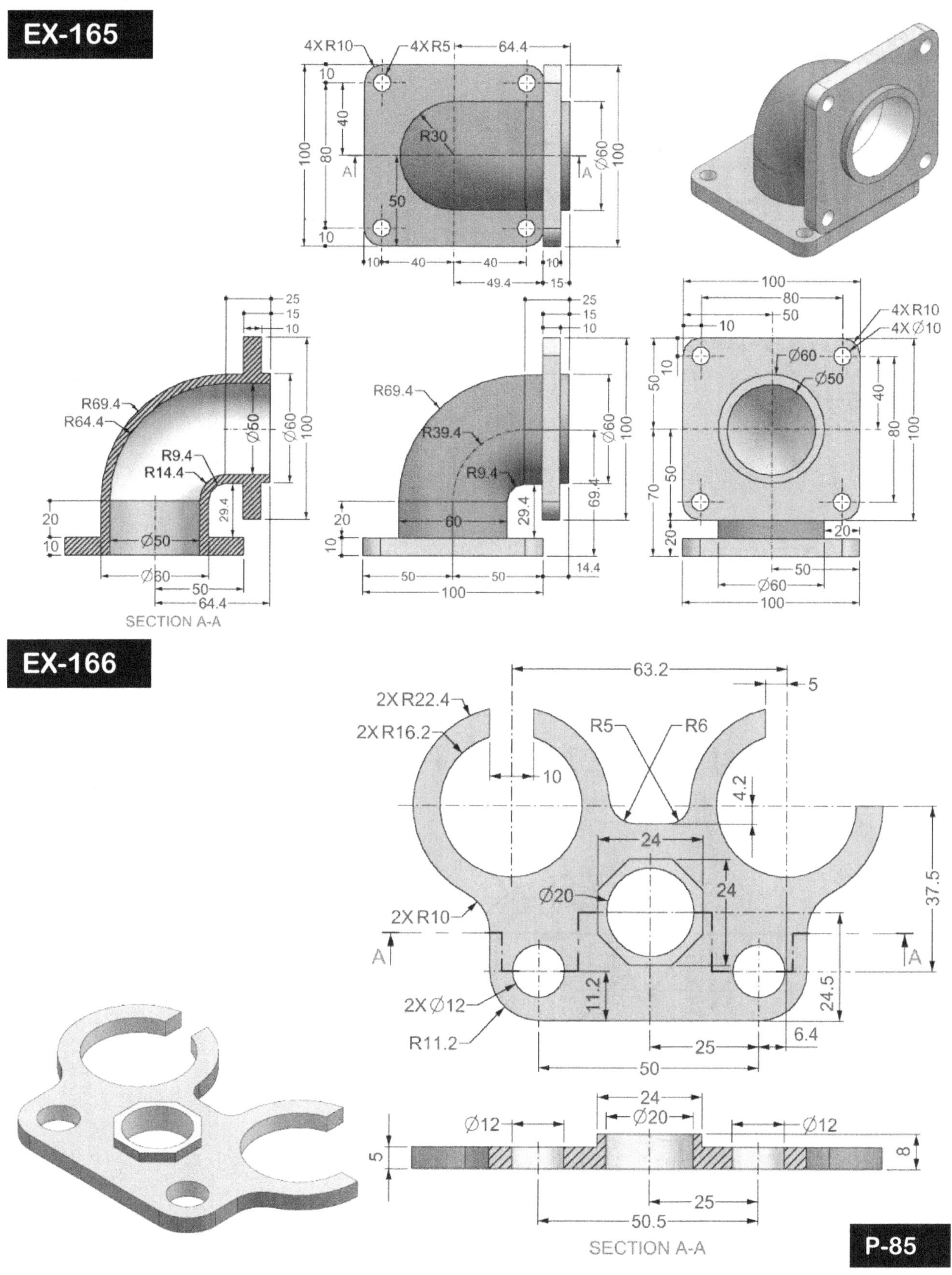

25
3
19
3
9.5
9.5
3.2
R5.6
Ø4.5
3
25
19
9.5
9.5
9.5
15.8
18.7
3
4X Ø5
Ø18
Ø15
R11.2
Ø16.2
5
15.8
25
5
Ø20.6
25
17
5
8
15.8
Ø22.4
25
5
Ø20.6
17
25
8
15.8
R11.2
15.8
Ø16.2
R10.3
R3
5
4X Ø5
15.8
2.9
Ø4.5
25
19
9.5
9.6
R5.6
R7.5
R3
12.6

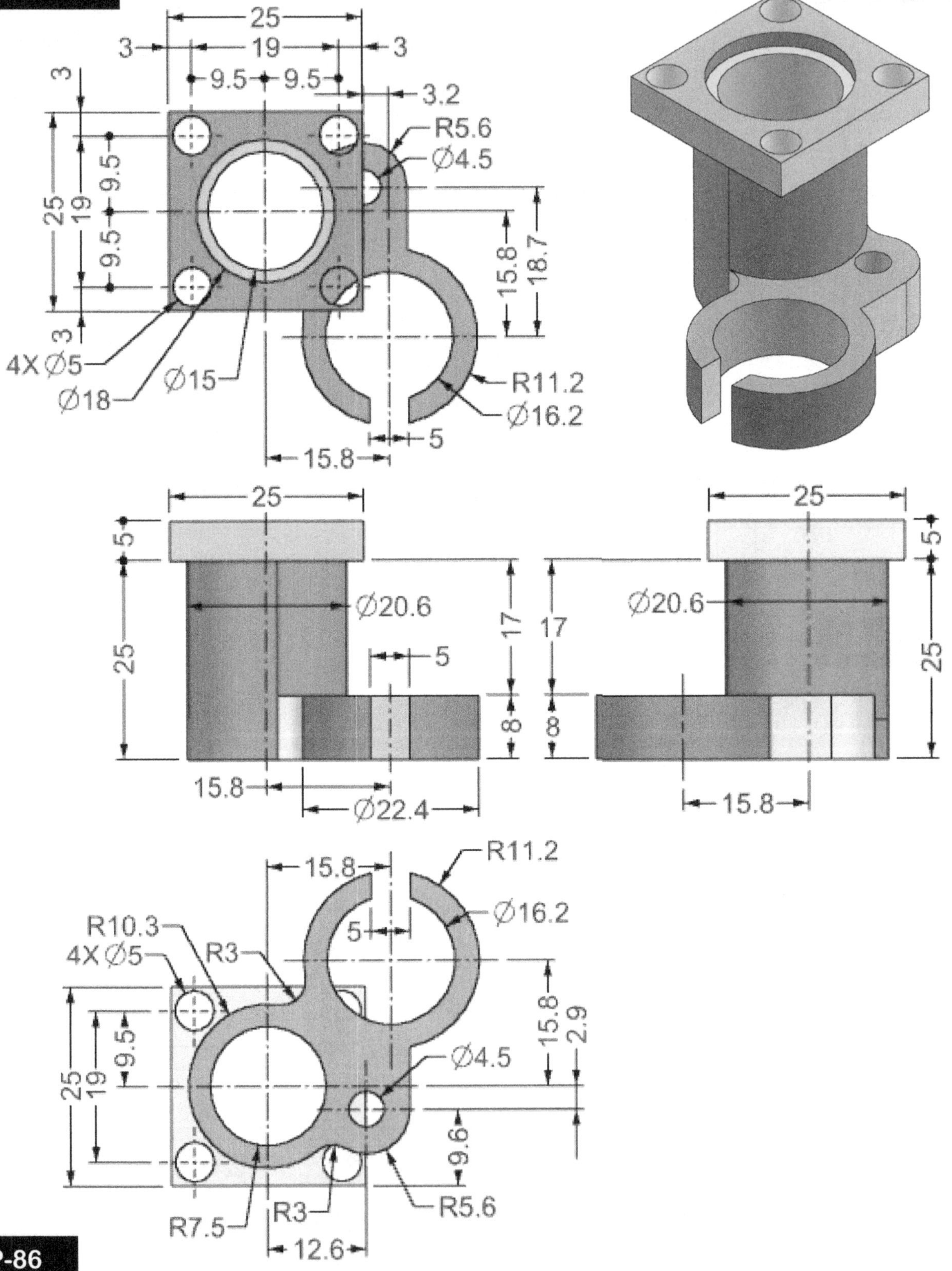

EX-168 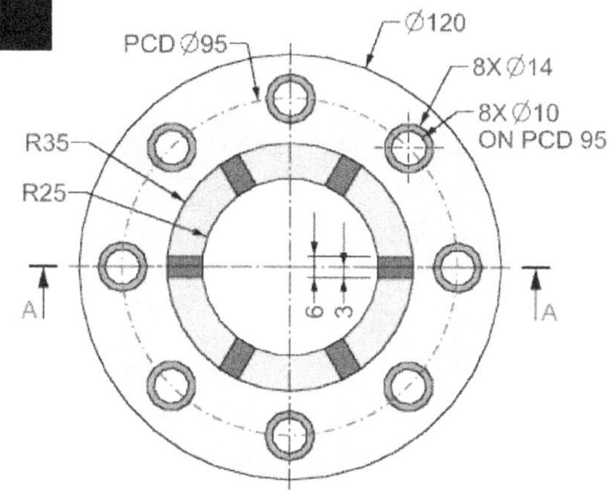 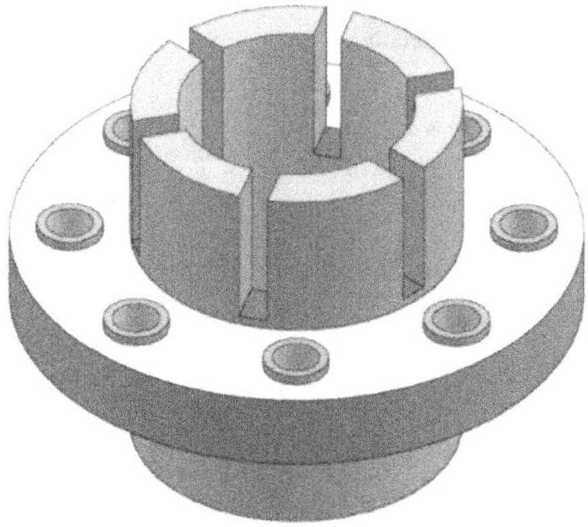 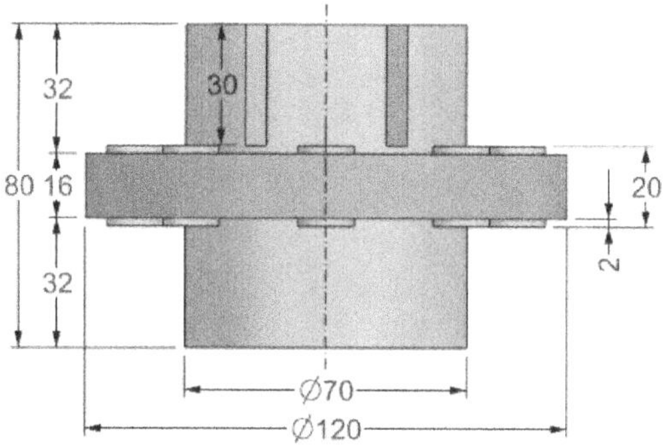 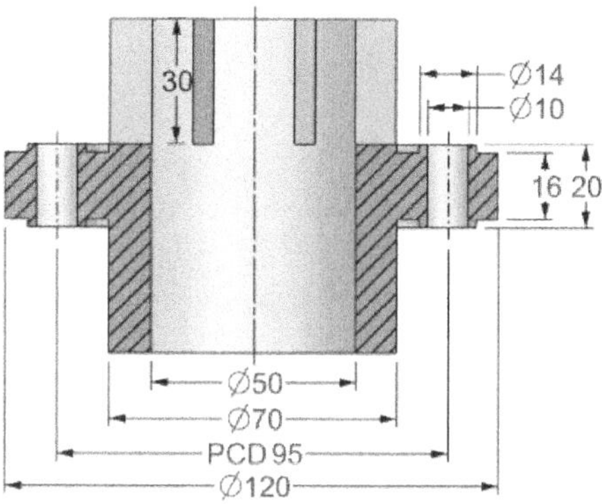
PCD Ø95
Ø120
8X Ø14
8X Ø10
ON PCD 95
R35
R25
A
A
6
3
32
30
80 16
20
32
2
Ø70
Ø120
30
Ø14
Ø10
16 20
Ø50
Ø70
PCD 95
Ø120
SECTION A-A
EX-169
Ø70
Ø40
20
R5
40
Ø28
Ø40
50
130
70
200
Ø70
R2
R60
Ø80
R5
30
40
21.3
50
80
Ø28
Ø40
140
30
10
50
80
70
Ø70
15
40
15
10
30
30
Ø28
Ø40
80
15
15
70
10
35
Ø70
P-87

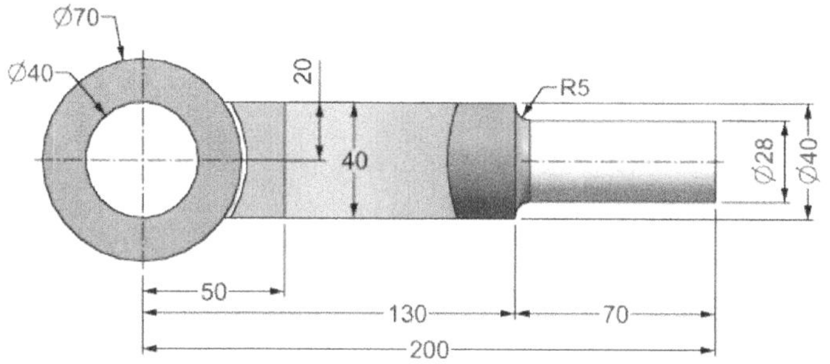

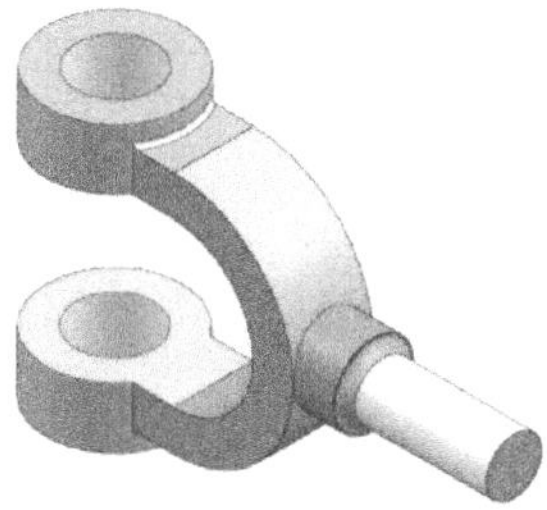

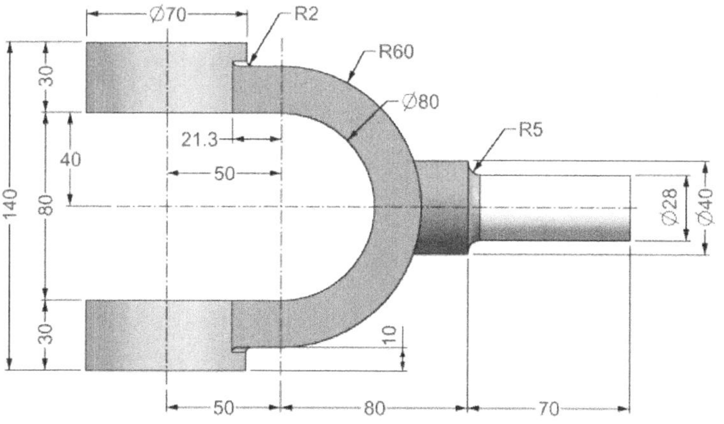

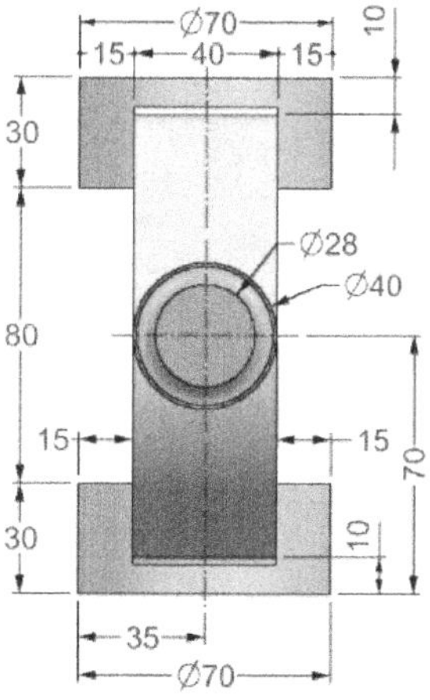

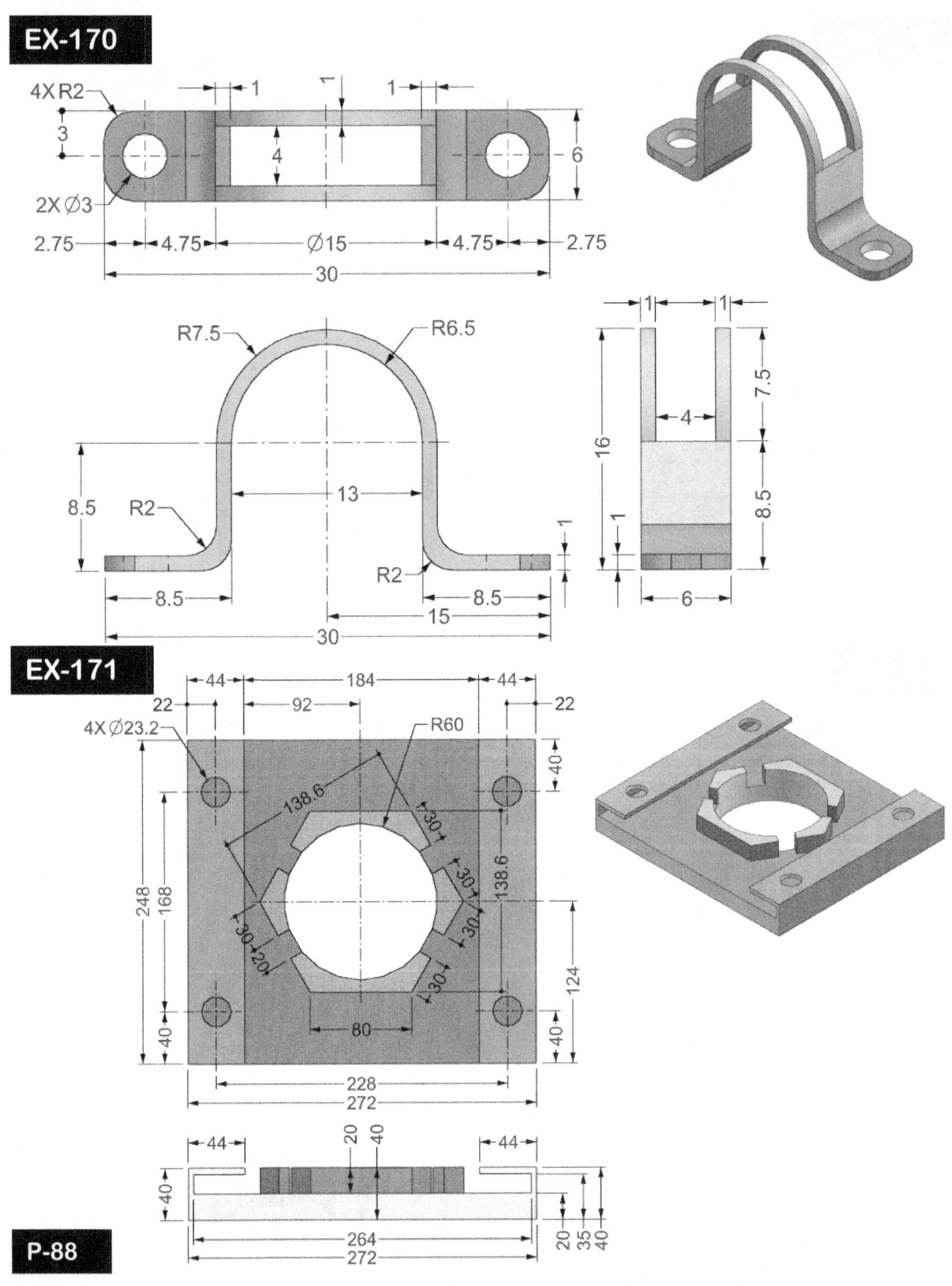

EX-170
4X R2
3
2X Ø3
1
1
1
4
6
2.75
4.75
Ø15
4.75
2.75
30
R7.5
R6.5
13
8.5
R2
R2
8.5
8.5
15
30
1
1
7.5
4
16
8.5
1
1
6
EX-171
44
184
44
22
92
22
4X Ø23.2
R60
138.6
30
30
40
248
168
138.6
30
30
30
20
30
124
80
40
40
228
272
44
20
40
44
40
264
20
35
40
272
P-88

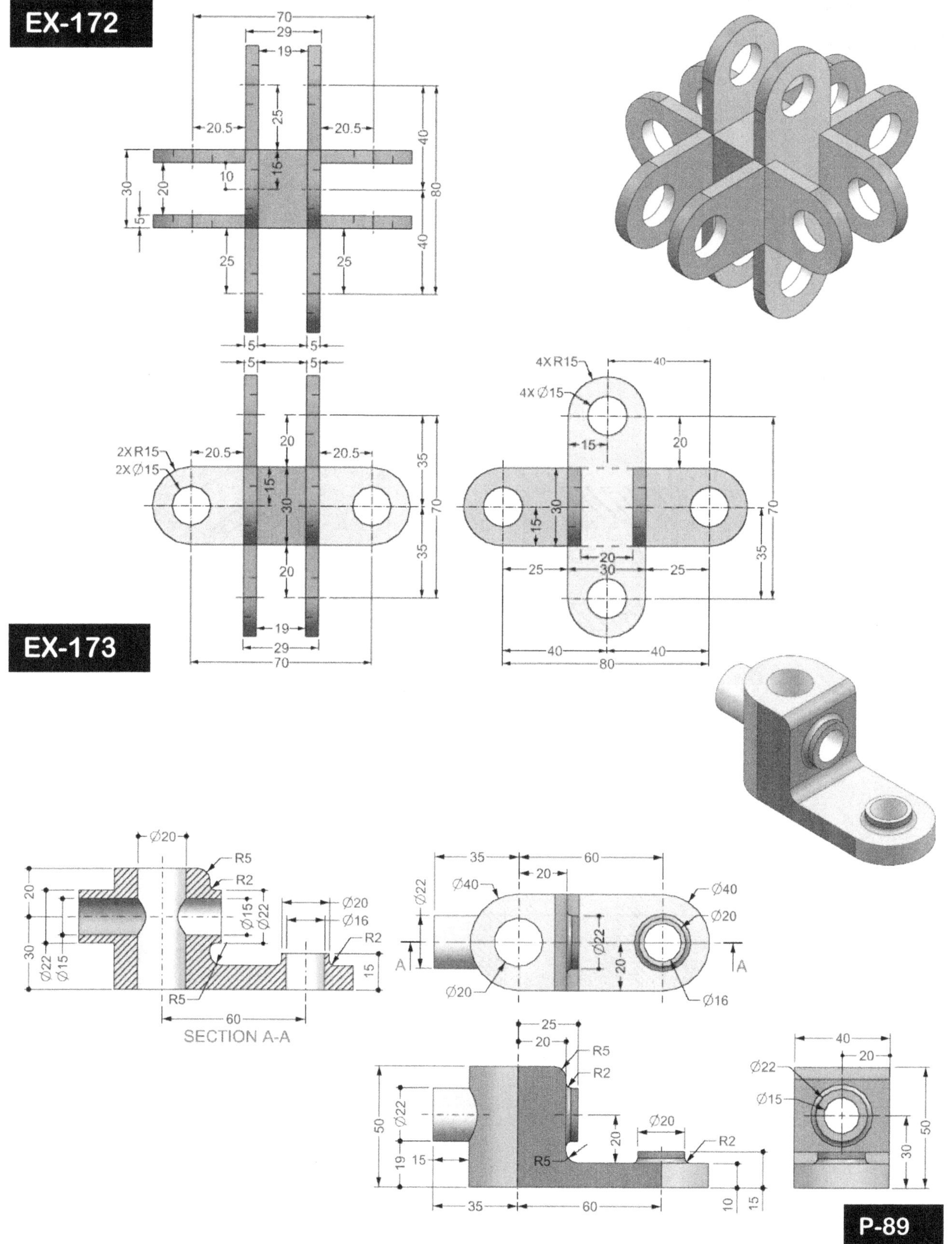

EX-172
EX-173
P-89
SECTION A-A
2X R15
2X Ø15
4X R15
4X Ø15
Ø20
Ø16
Ø15
Ø22
Ø40
R5
R2
A

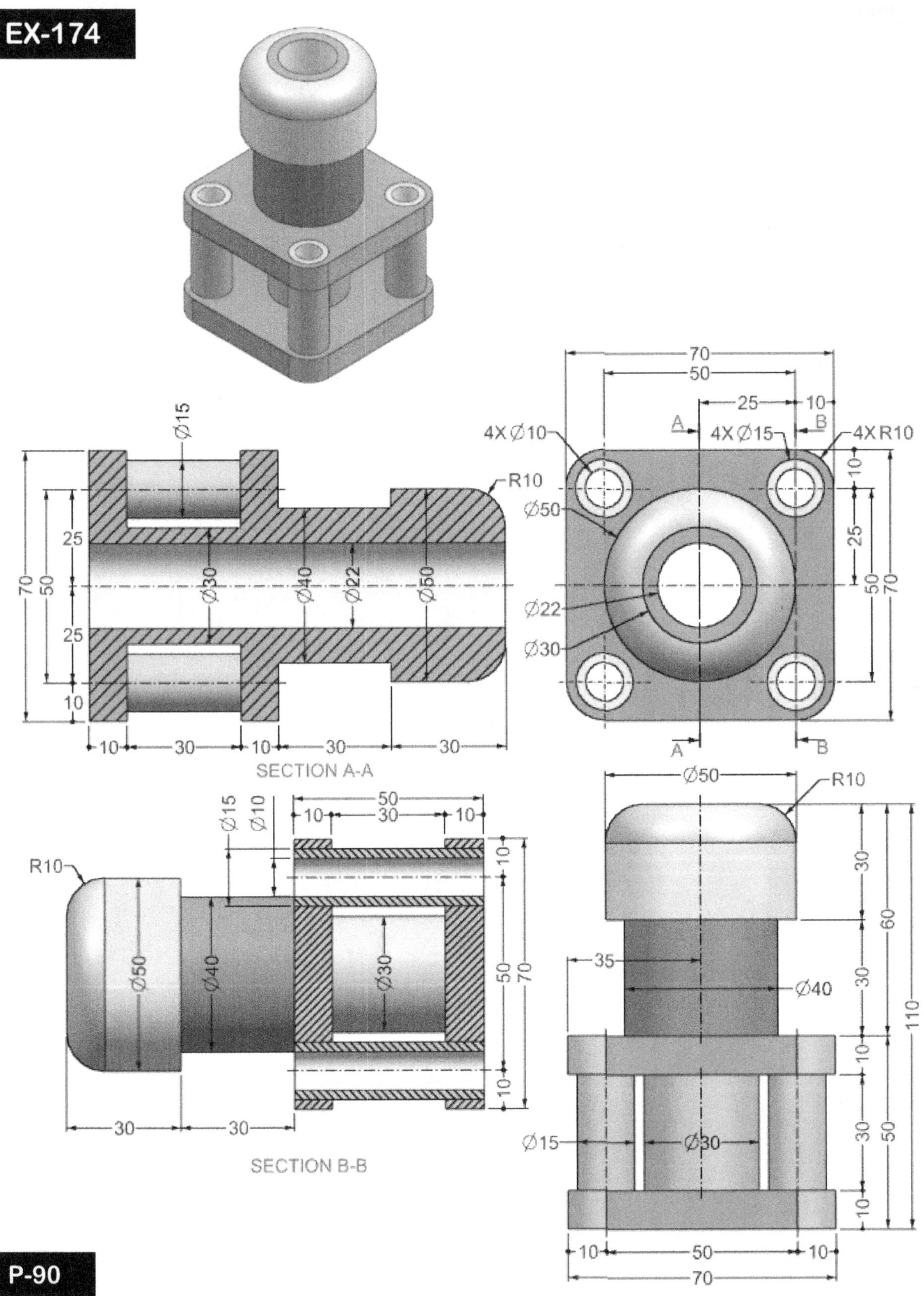

EX-174
P-90
Ø15
SECTION A-A
70
50
25
10
4X Ø10
4X Ø15
4X R10
R10
Ø50
25
50
70
10
Ø22
Ø30
A
B
25
70
50
25
10
Ø30
Ø40
Ø22
Ø50
10
30
10
30
30
Ø15
Ø10
50
10
30
10
R10
Ø50
Ø40
Ø30
10
50
70
10
SECTION B-B
30
30
Ø50
R10
30
60
35
Ø40
110
10
30
Ø15
Ø30
50
30
10
10
50
10
70

70
35
35
10
25
25
10
4X R10
4X Ø15
4X Ø10
Ø30
Ø22
10
35
25
70
A
A
25
35
10
Ø30
Ø15
Ø15
20
5
10
Ø15
Ø15
30
50
50
10
70
Ø30
Ø22
Ø15
Ø10
20
5
10
30
70
30
Ø22
25
25
50
70
10
SECTION A-A

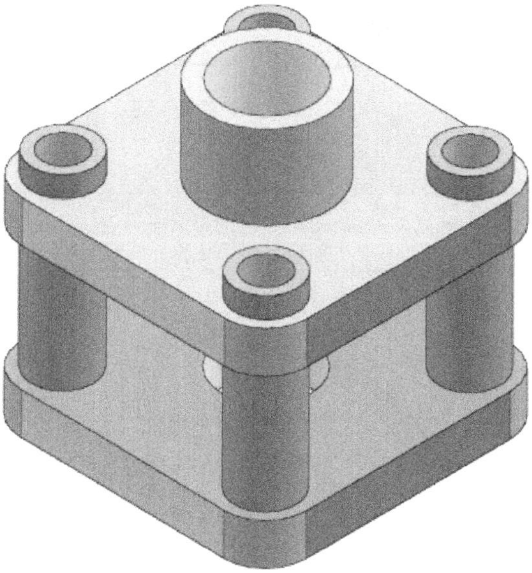

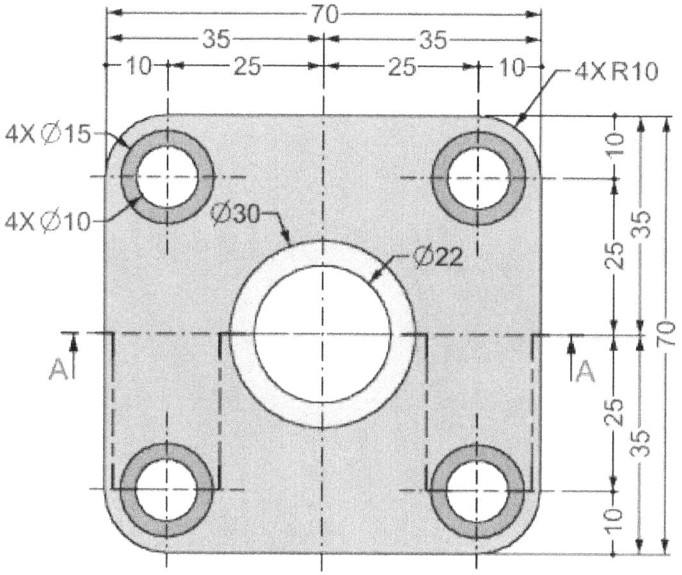

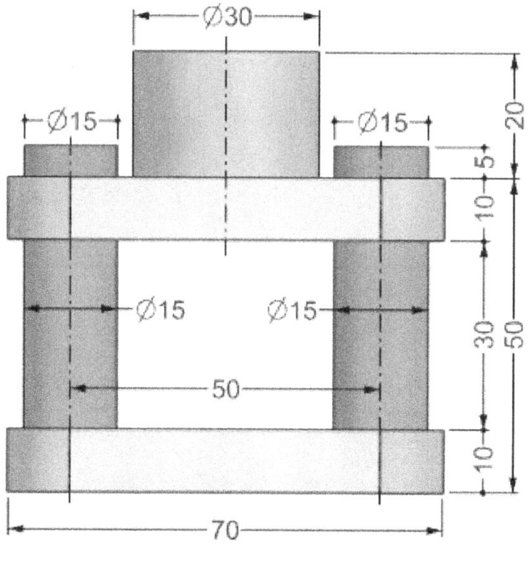

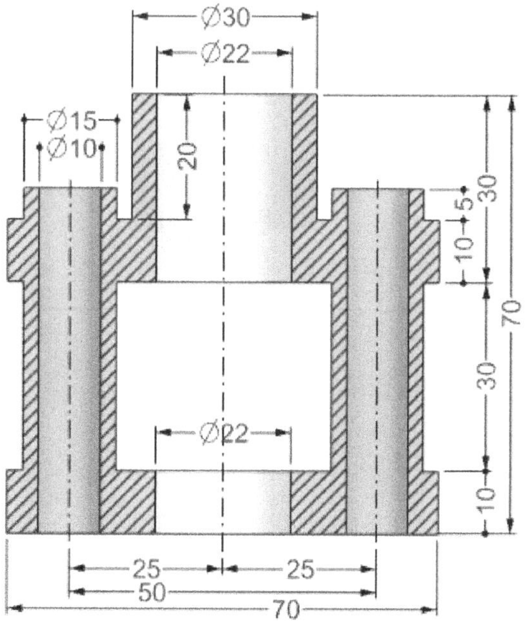

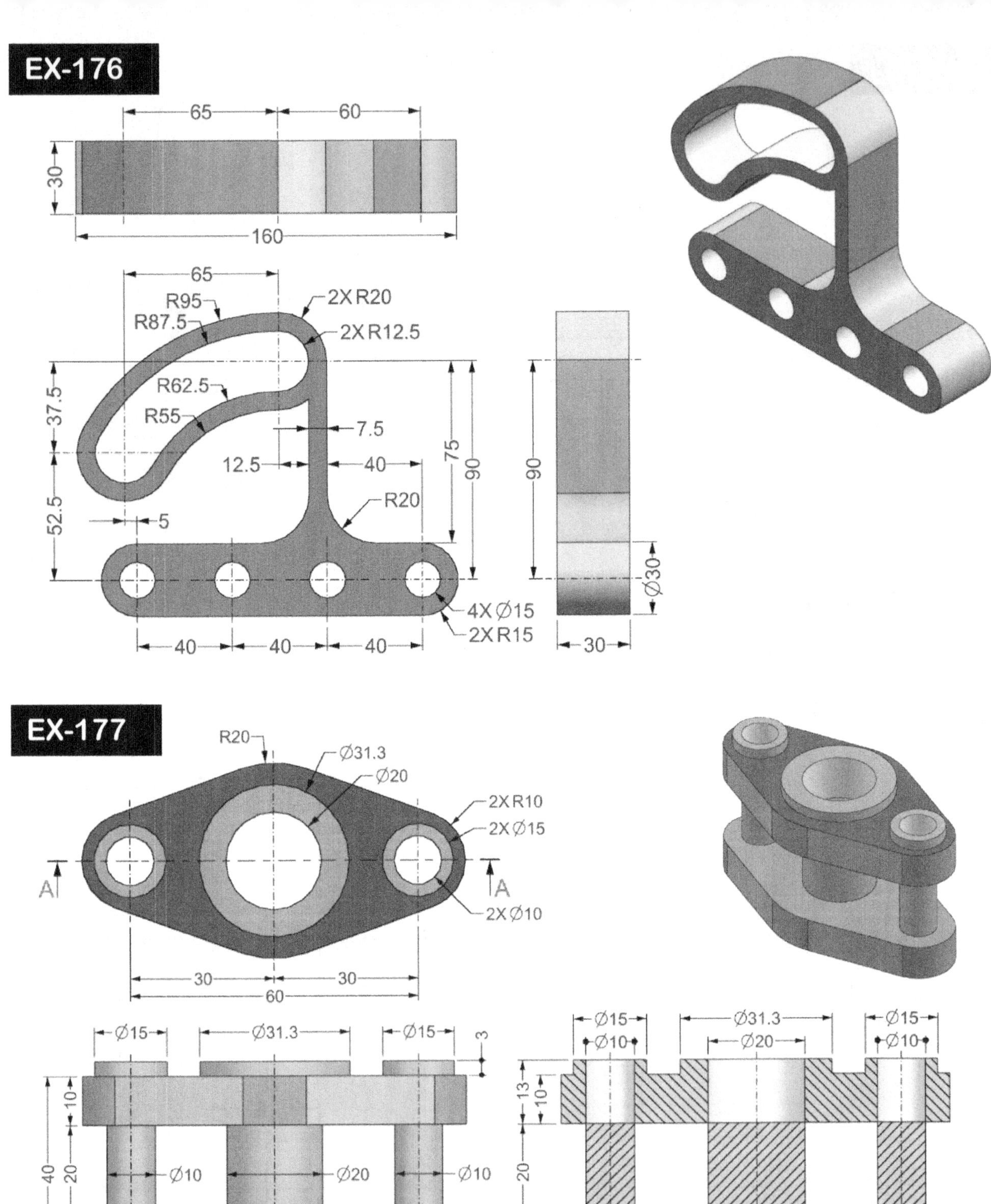

EX-176
65
60
30
160
65
R95
2X R20
R87.5
2X R12.5
37.5
R62.5
R55
7.5
75
90
90
52.5
12.5
40
5
R20
Ø30
4X Ø15
40
40
40
2X R15
30

EX-177
R20
Ø31.3
Ø20
2X R10
2X Ø15
A
A
2X Ø10
30
30
60
Ø15
Ø31.3
Ø15
3
Ø15
Ø31.3
Ø15
Ø10
Ø10
10
Ø10
Ø20
Ø10
13
10
40
20
20
10
10
30
30
30
30
60
60
SECTION A-A

P-92

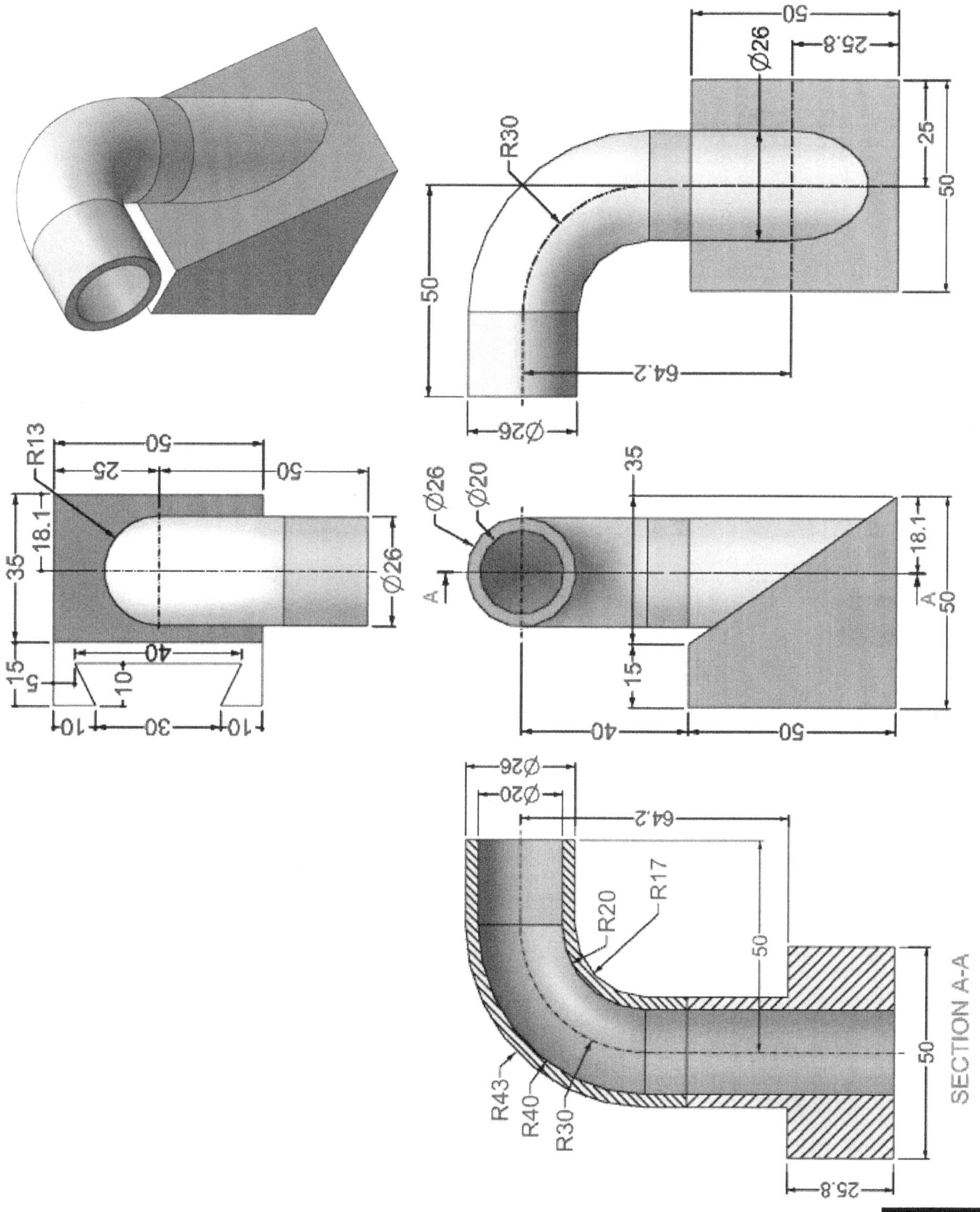
R30
Ø26
50
25.8
25
50
64.2
Ø26
R13
50
25
50
35
18.1
Ø26
40
15
5
10
10
30
Ø26
Ø20
35
A
A
18.1
50
15
40
50
Ø26
Ø20
64.2
R20
R17
50
R43
R40
R30
50
25.8
SECTION A-A

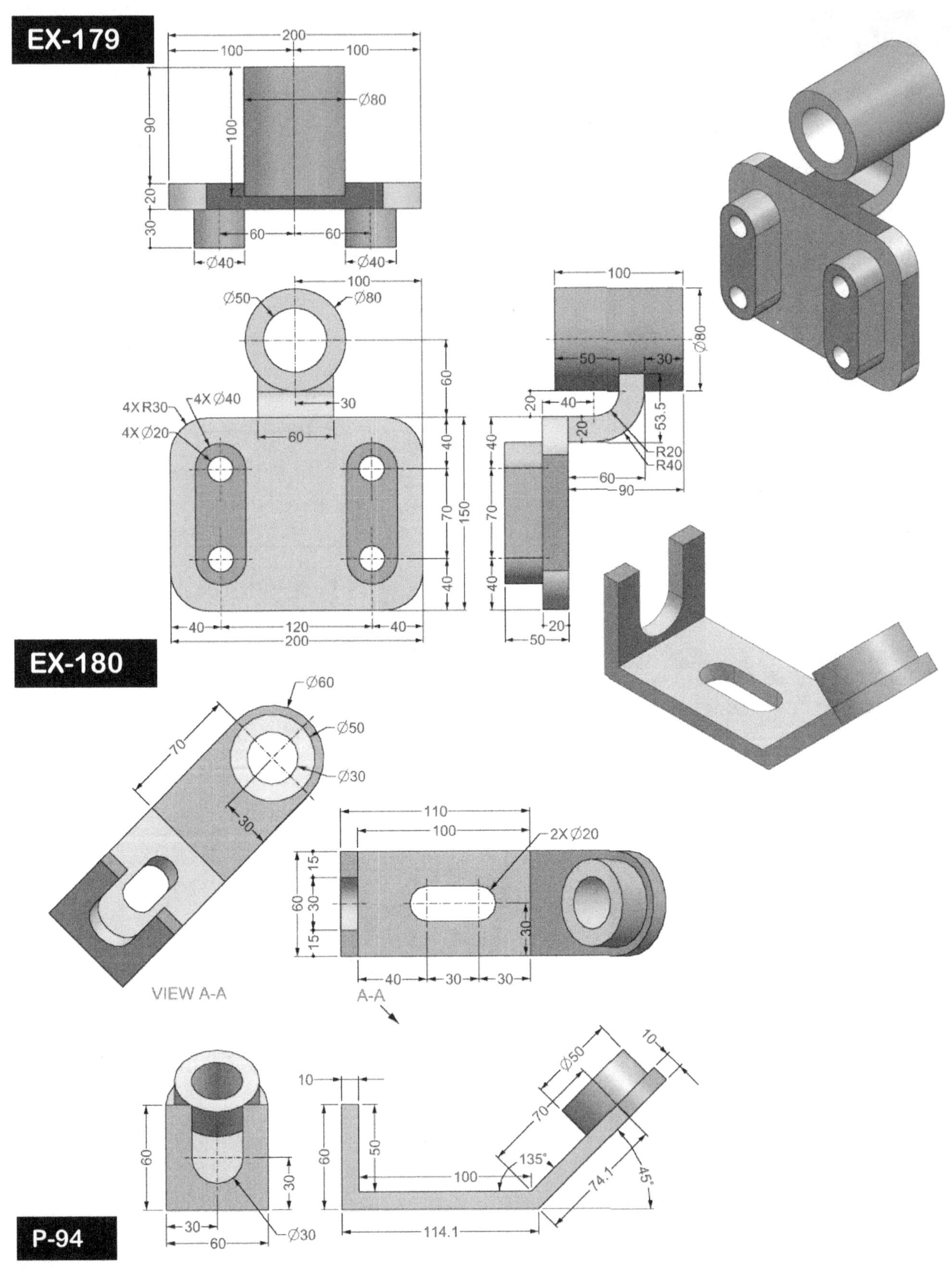

EX-179
EX-180
P-94
VIEW A-A
A-A
200
100
100
90
100
20
30
Ø80
60
60
Ø40
Ø40
Ø50
Ø80
4X R30
4X Ø40
4X Ø20
30
60
60
40
150
70
40
40
120
40
200
100
50
30
Ø80
20
40
20
53.5
R20
R40
60
90
40
70
40
20
50
Ø60
Ø50
Ø30
70
30
110
100
2X Ø20
15
30
15
60
30
40
30
30
10
60
50
Ø50
70
135°
100
45°
74.1
114.1
60
30
30
60
Ø30

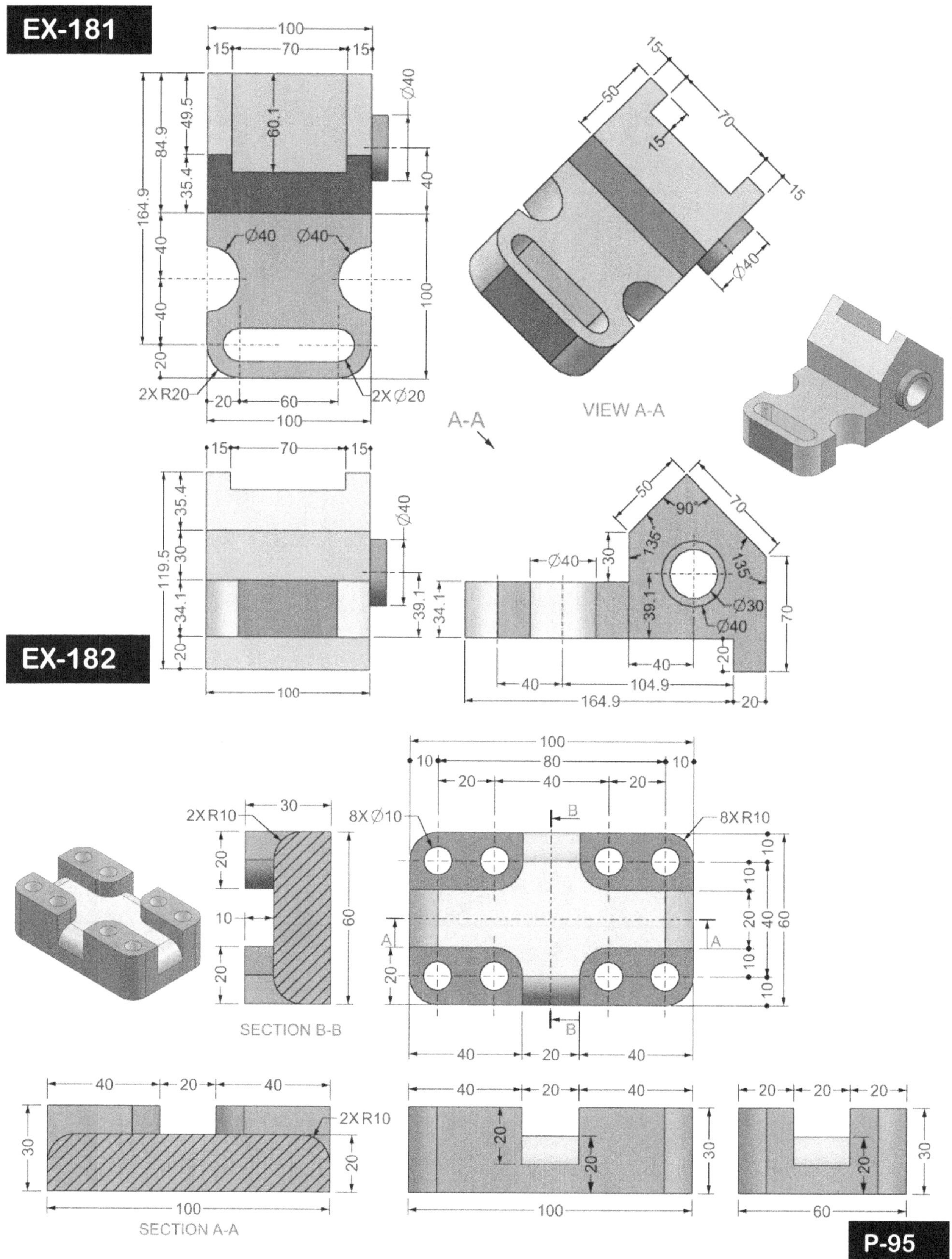

EX-181
100
15
70
15
Ø40
60.1
49.5
84.9
35.4
164.9
40
Ø40
Ø40
100
40
40
20
2X R20
20
60
100
2X Ø20
A-A
VIEW A-A
15
50
70
15
15
Ø40
EX-182
15
70
15
35.4
Ø40
30
119.5
39.1
34.1
20
100
Ø40
30
50
90°
135°
135°
70
39.1
34.1
Ø30
Ø40
40
20
40
104.9
20
164.9
70
100
10
80
10
20
40
20
2X R10
30
8X Ø10
B
8X R10
20
10
10
60
10
20
40
60
A
A
10
20
40
SECTION B-B
B
40
20
40
40
20
40
20
20
20
40
20
40
30
2X R10
20
20
30
20
20
30
30
100
100
60
SECTION A-A
P-95

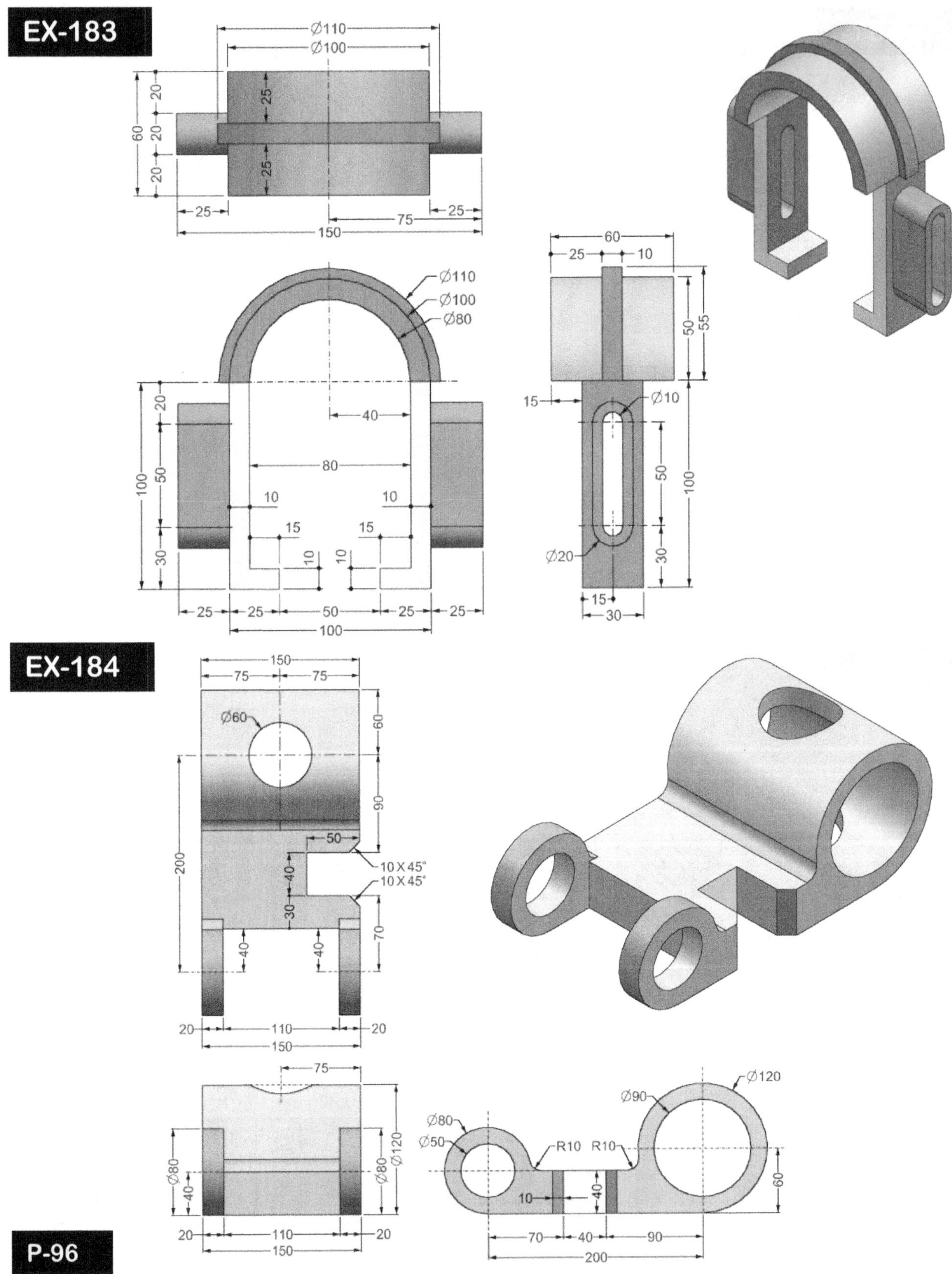

EX-183
EX-184
P-96

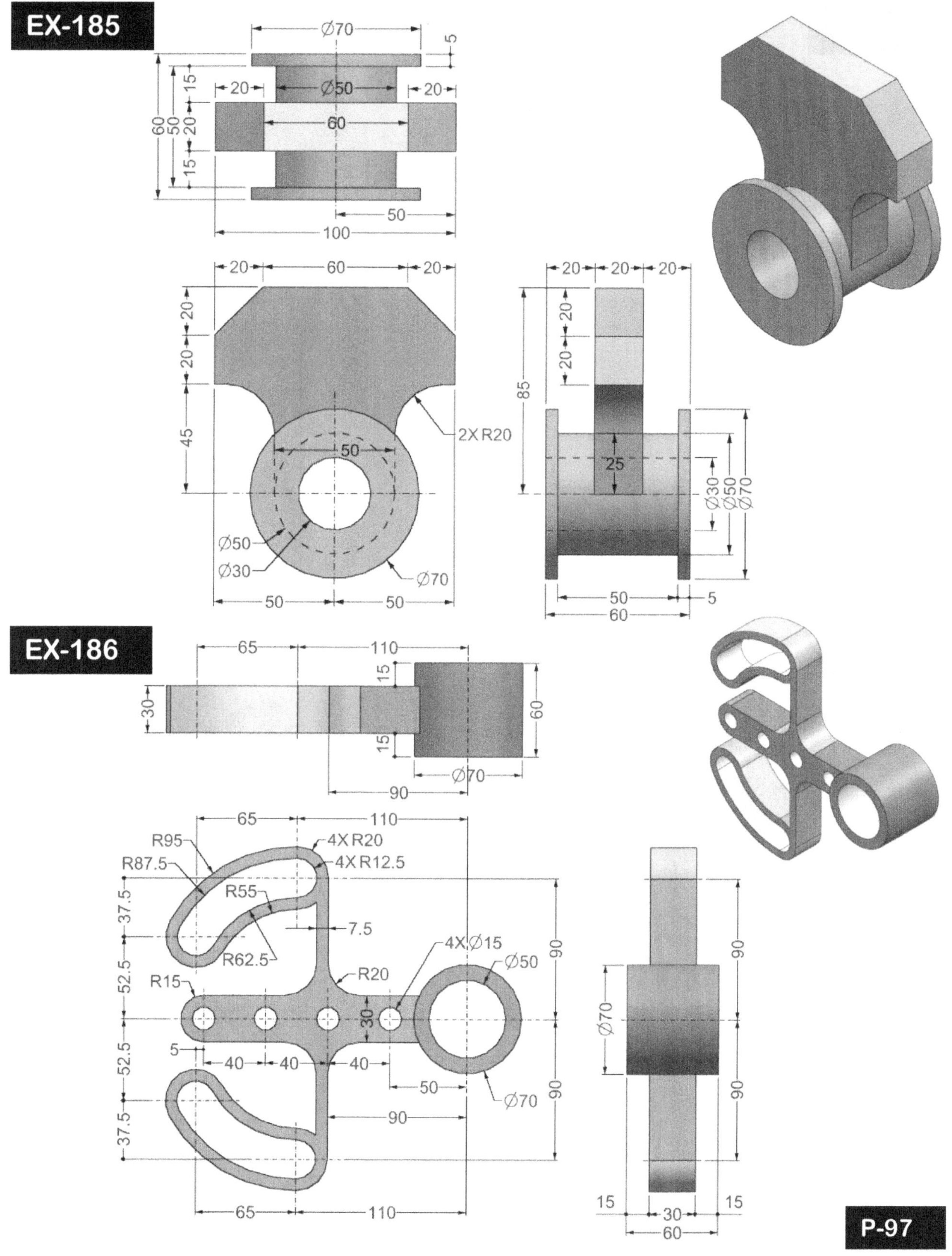

EX-185
EX-186
P-97

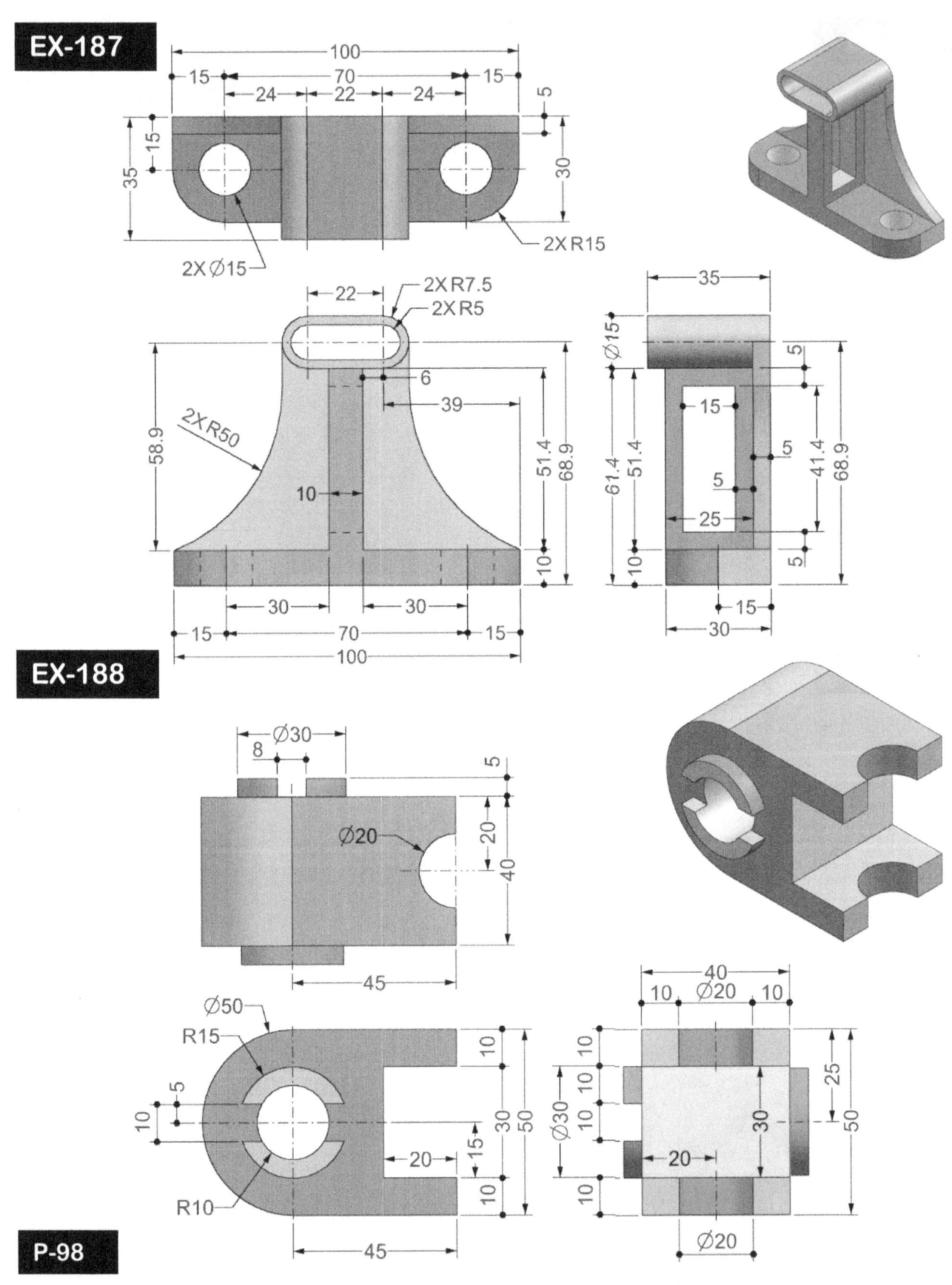

EX-187
100
15
70
15
24
22
24
5
15
35
30
2X R15
2X Ø15
22
2X R7.5
2X R5
6
39
2X R50
58.9
51.4
68.9
10
10
35
Ø15
5
15
61.4
51.4
5
5
41.4
68.9
25
10
5
15
30
30
30
15
70
15
100
EX-188
Ø30
8
5
Ø20
20
40
45
40
10
Ø20
10
Ø50
10
R15
30
10
10
5
Ø30
10
25
10
30
50
20
50
15
30
20
R10
10
45
Ø20
P-98

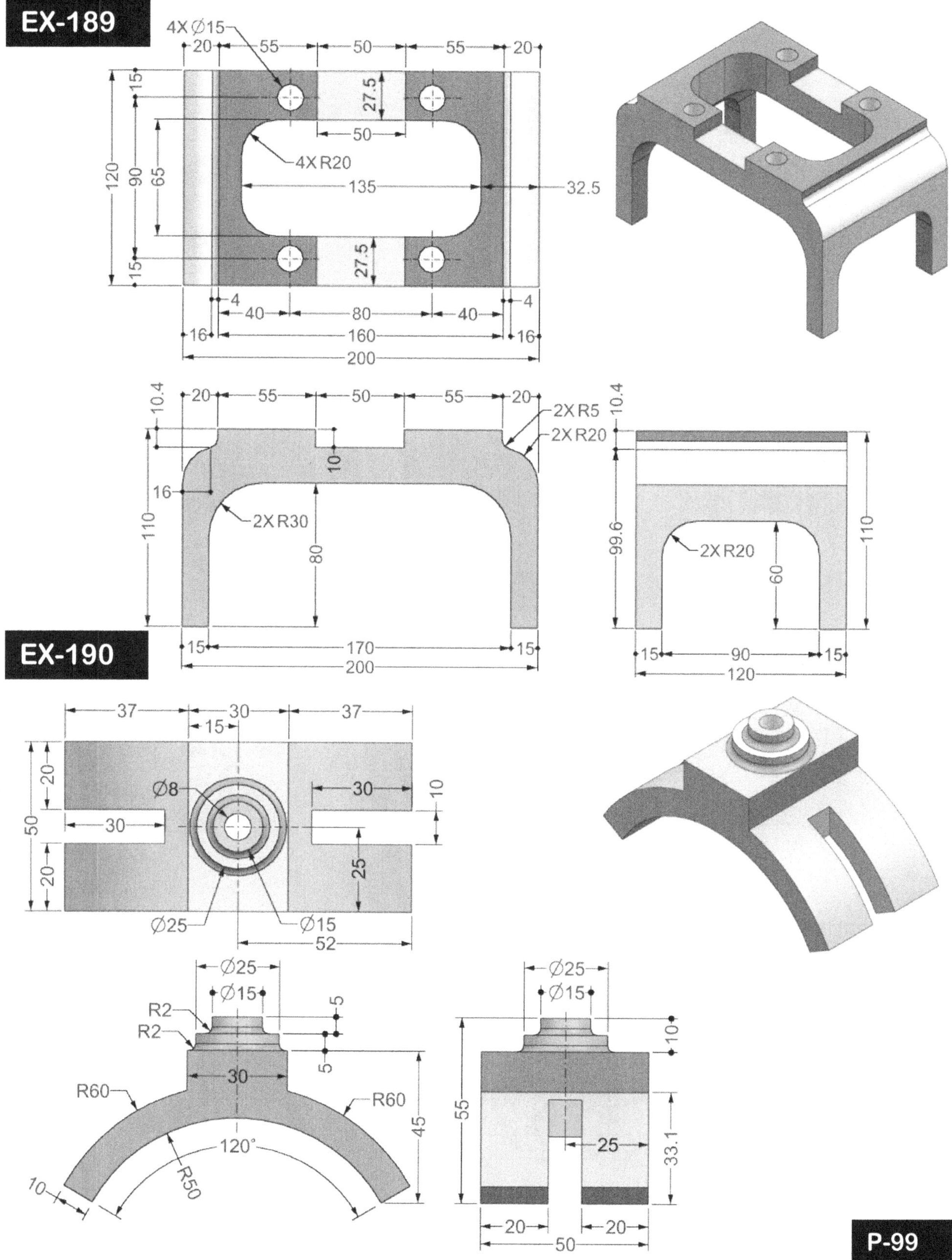

EX-189
4X Ø15
4X R20
EX-190
2X R5
2X R20
2X R30
2X R20
Ø8
Ø25
Ø15
R2
R2
R60
R60
120°
R50
P-99

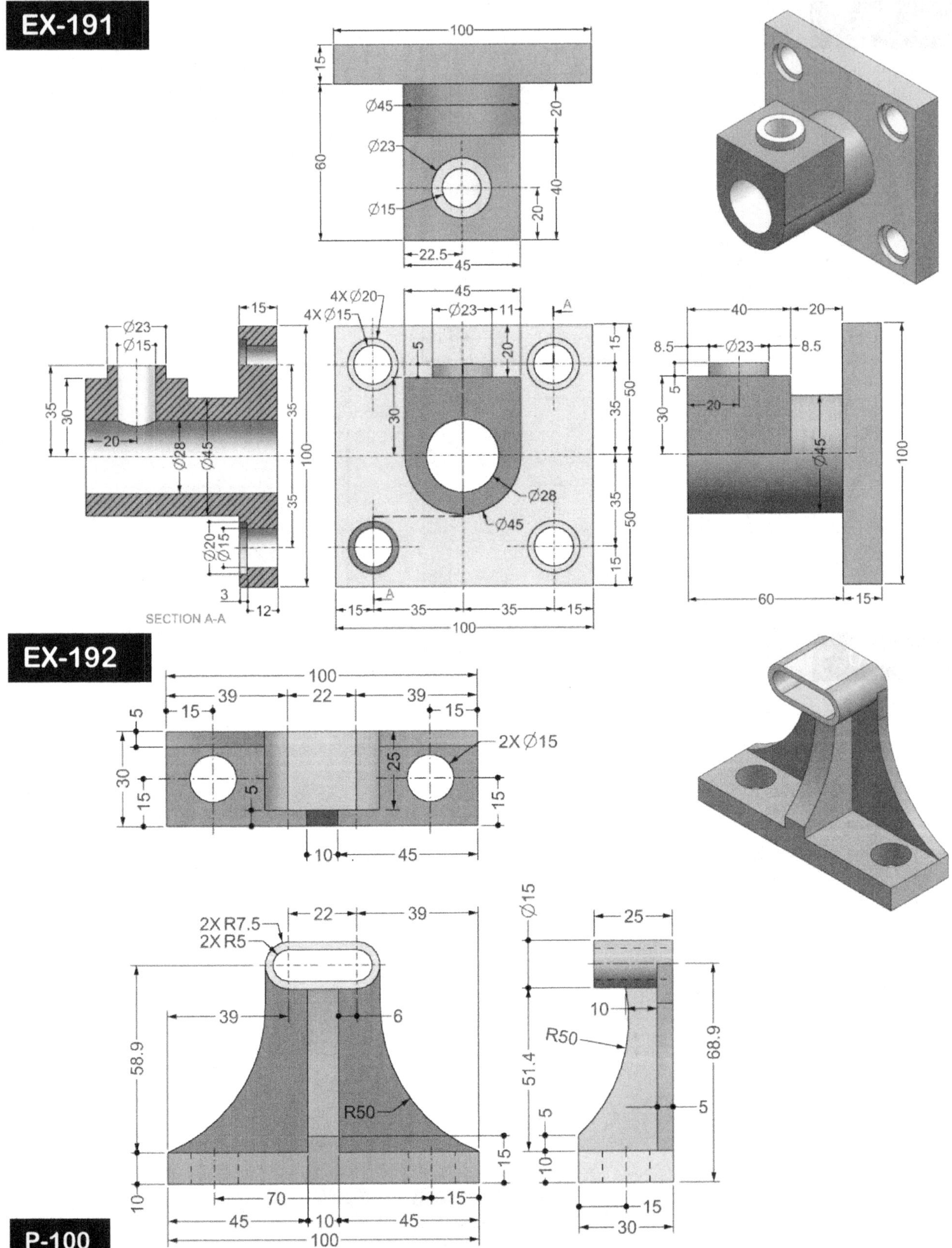

EX-191
EX-192
P-100
SECTION A-A

EX-193

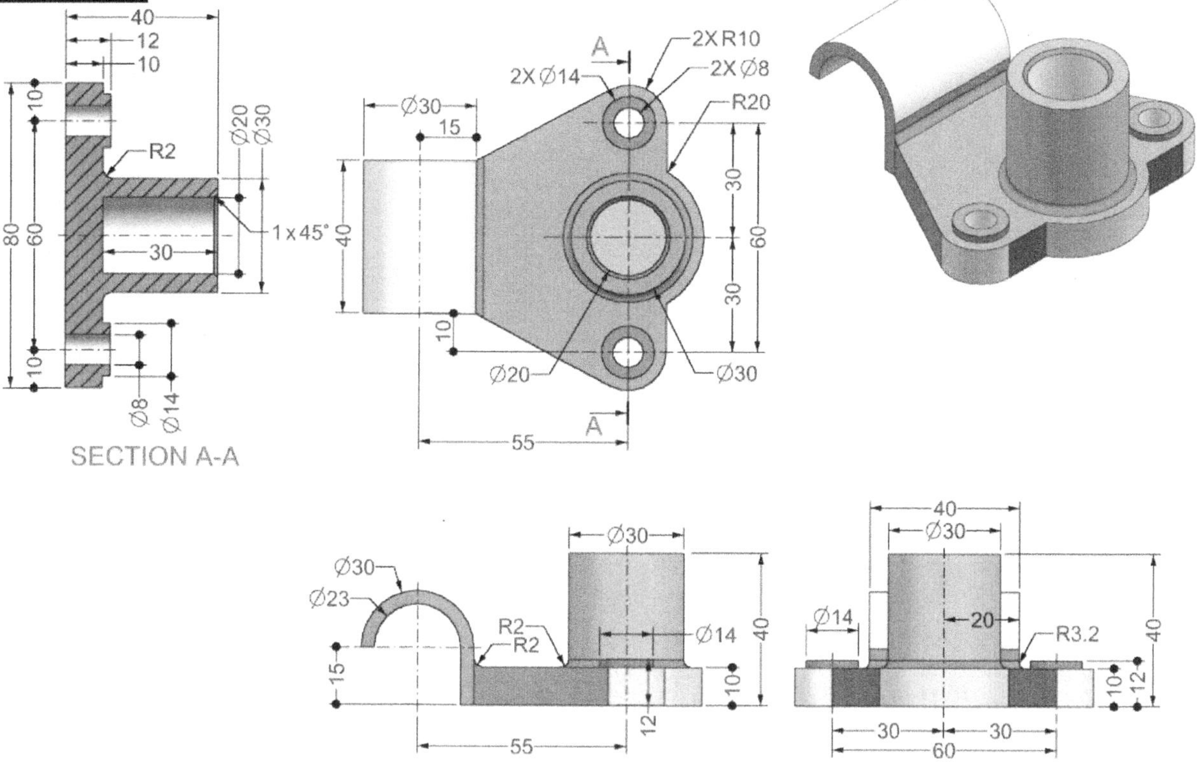

40
12
10
10
Ø20
Ø30
R2
80
60
30
1 x 45°
10
Ø8
Ø14
SECTION A-A

A
2X R10
2X Ø14
2X Ø8
R20
Ø30
15
40
30
60
30
10
Ø20
Ø30
55
A

Ø30
Ø23
R2
R2
Ø14
15
Ø30
40
12
55
10

40
Ø30
Ø14
20
R3.2
30
30
60
10
12
40

EX-194

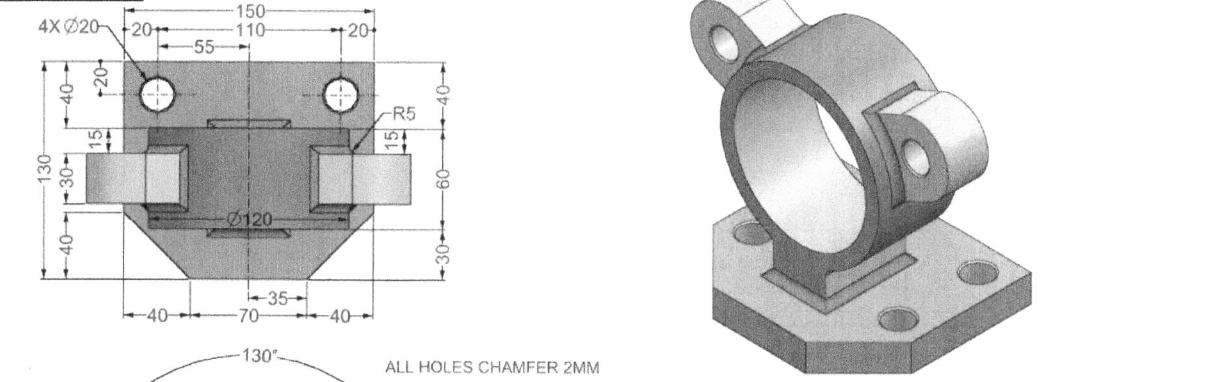

150
110
4X Ø20
20
20
55
20
40
40
15
15
130
30
R5
60
40
Ø120
30
40
70 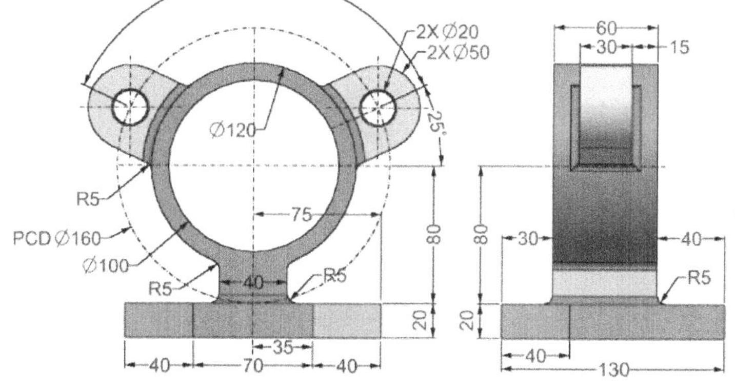
40
35

130°
ALL HOLES CHAMFER 2MM
2X Ø20
2X Ø50
25°
Ø120
R5
75
80
PCD Ø160
Ø100
R5
R5
40
20
40
70
40
35

60
30
15
80
30
40
20
R5
40
130
R5

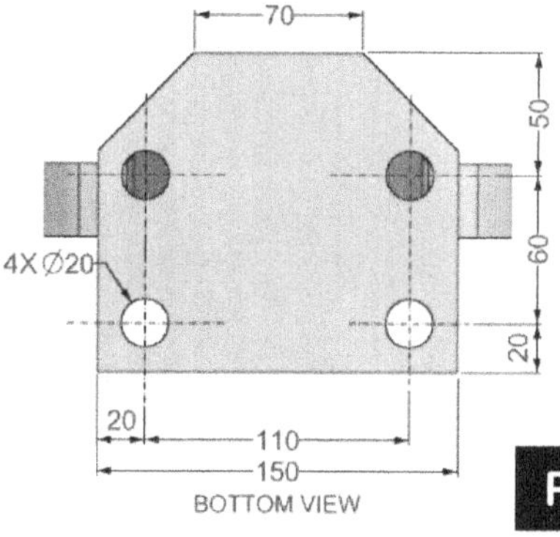

70
50
60
4X Ø20
20
20
110
150
BOTTOM VIEW

P-101

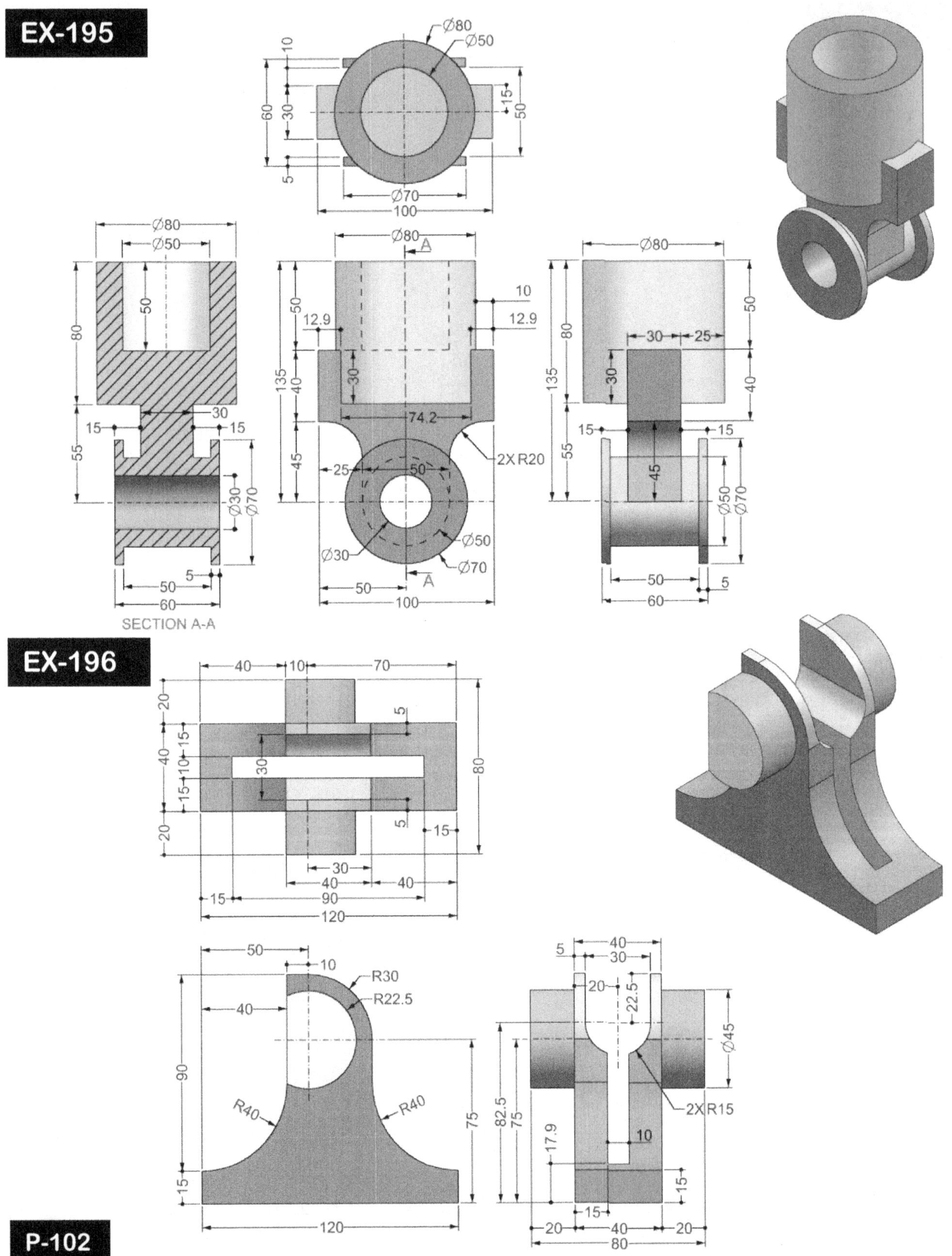

EX-195
SECTION A-A
EX-196
P-102
Ø80
Ø50
Ø70
100
10
30
60
5
15
50
Ø80
Ø50
50
80
30
15
15
55
Ø30
Ø70
50
5
60
Ø80
A
50
135
40
45
12.9
30
74.2
25
50
Ø30
Ø50
Ø70
50
100
2X R20
A
10
12.9
Ø80
80
30
25
135
55
30
45
15
15
Ø50
Ø70
50
5
60
40
10
70
20
40
15
10
15
20
30
80
5
5
15
15
30
40
40
90
120
50
10
40
R30
R22.5
R40
R40
90
75
15
120
5
40
30
20
22.5
Ø45
2X R15
82.5
75
17.9
10
15
20
40
20
80
15

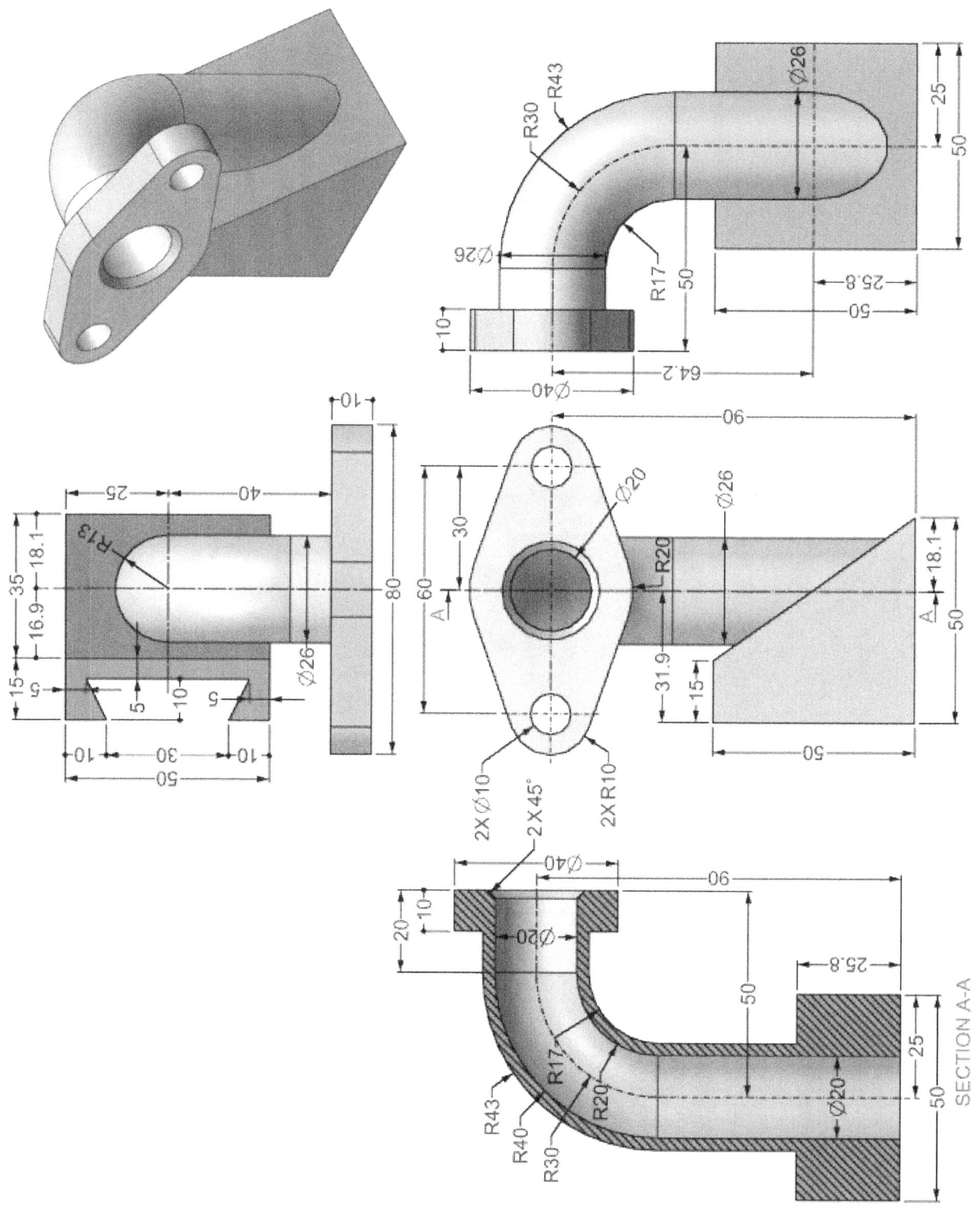

R43
R30
Ø26
25
50
Ø26
R17
50
25.8
50
10
Ø40
64.2
90
Ø20
Ø26
30
R20
60
A
A
31.9
15
18.1
50
50
10
25
40
R13
35
18.1
16.9
15
5
5
10
5
Ø26
80
10
30
50
2X Ø10
2X 45°
2X R10
Ø40
20
10
Ø20
90
50
25.8
R17
R20
R43
R40
R30
25
Ø20
50
SECTION A-A

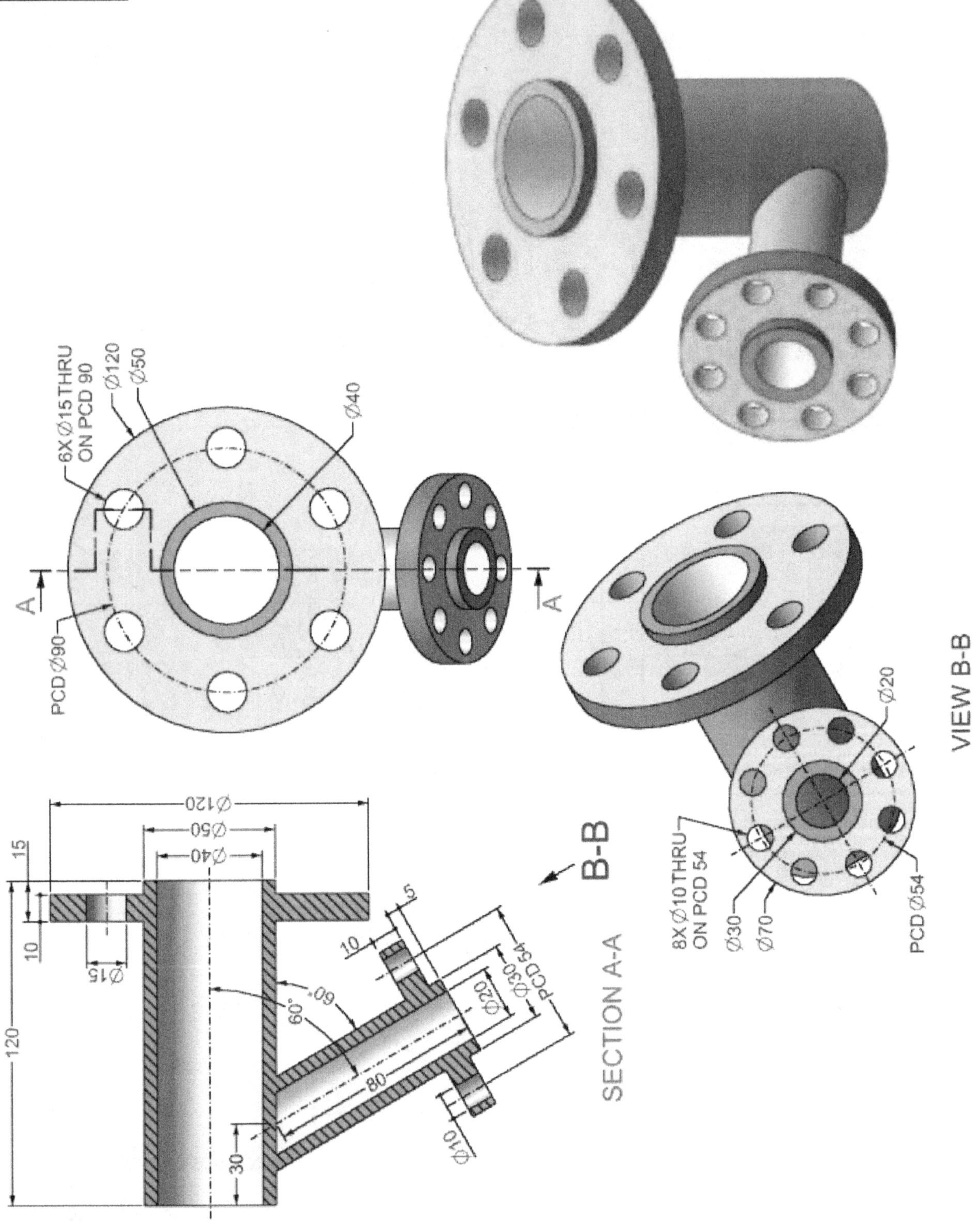

EX-198

6X Ø15 THRU ON PCD 90
Ø120
Ø50
Ø40
PCD Ø90

A
A

VIEW B-B

Ø20
8X Ø10 THRU ON PCD 54
Ø30
Ø70
PCD Ø54

B-B

SECTION A-A

Ø120
Ø50
Ø40
15
10
Ø15
120
60°
60°
80
30
Ø10
5
10
Ø20
Ø30
PCD 54

P-104

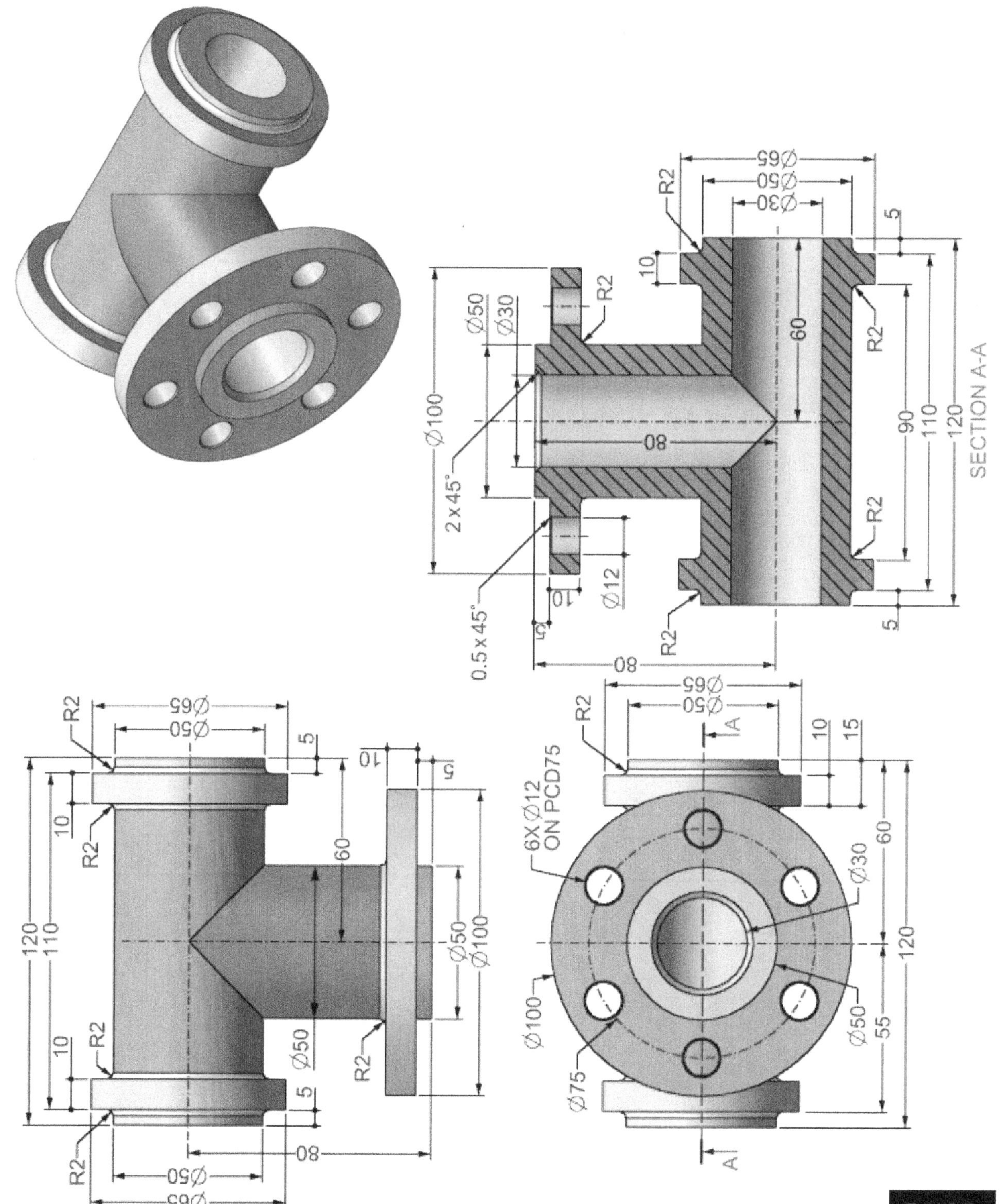

R2
Ø65
Ø50
Ø30
5
10
60
90
110
120
SECTION A-A
R2
R2
R2
Ø50
Ø30
Ø100
2 x 45°
0.5 x 45°
Ø12
10
5
80
80
R2
R2
Ø65
Ø50
A
10
15
6X Ø12
ON PCD75
R2
Ø30
60
120
Ø100
Ø50
55
Ø75
A
R2
Ø65
Ø50
5
10
10
60
Ø50
Ø100
R2
R2
R2
120
110
10
5
80
Ø50
Ø65

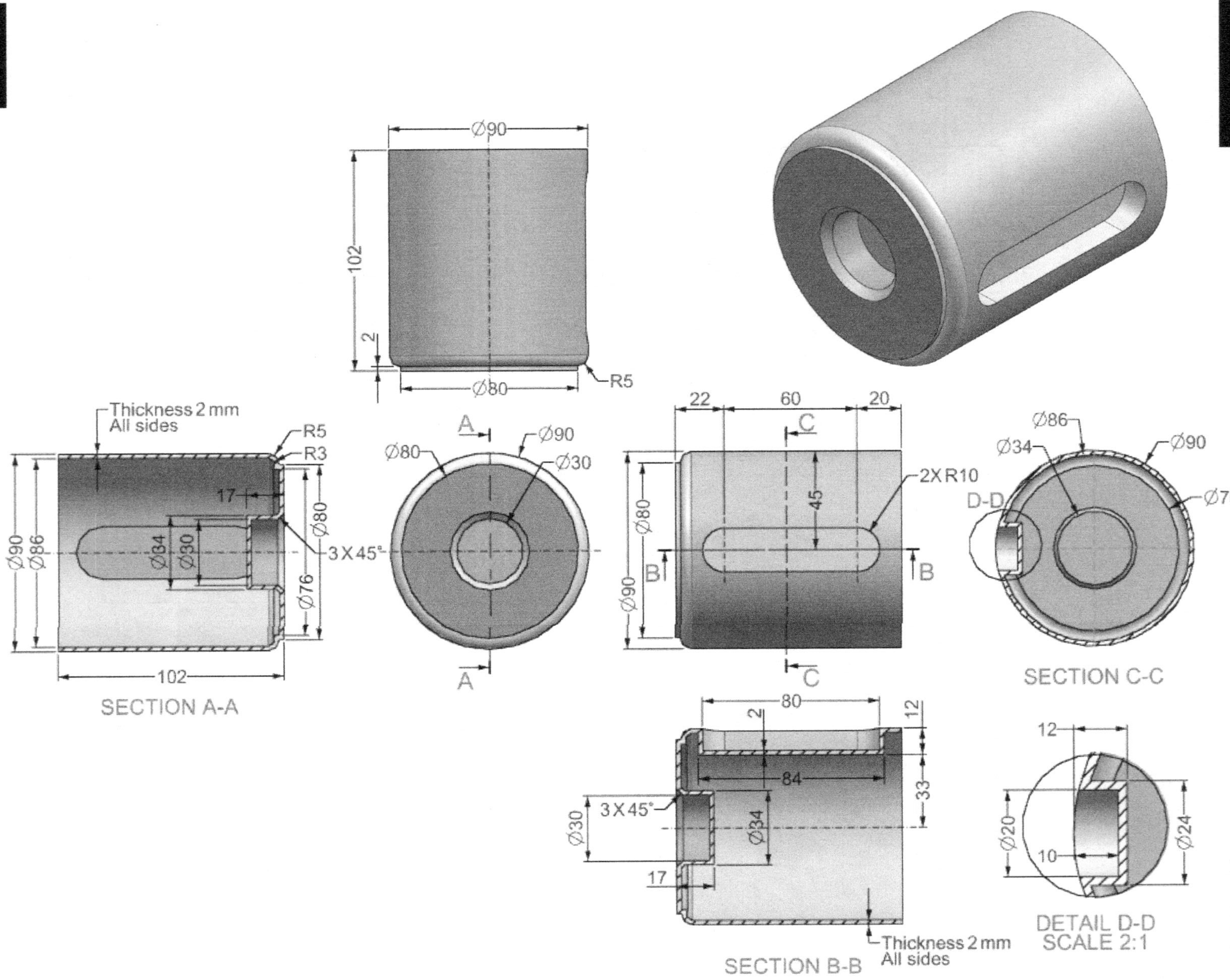

Ø90
102
2
Ø80
R5
Thickness 2 mm
All sides
R5
R3
17
Ø34
Ø30
3 X 45°
Ø90
Ø86
Ø76
Ø80
102
SECTION A-A
A
A
Ø80
Ø90
Ø30
22
60
20
C
45
Ø80
Ø90
B
B
2X R10
C
D-D
Ø86
Ø34
Ø90
Ø76
SECTION C-C
80
2
12
84
33
Ø30
3 X 45°
Ø34
17
Thickness 2 mm
All sides
SECTION B-B
12
Ø20
10
Ø24
DETAIL D-D
SCALE 2:1

Other useful books by CADIN360

1. 150 CAD Exercises

2. AutoCAD Exercises

3. CAD Exercises

4. 50+ SolidWorks Exercises

5. SolidWorks 200 Exercises

6. Autodesk Inventor Exercises

7. Catia Exercises

8. Siemens NX Exercises

Made in the USA
Monee, IL
09 July 2026